IGNITION

Also by Maura O'Connor:

*Resurrection Science: Conservation,
De-Extinction and the Precarious Future of Wild Things*

*Wayfinding: The Science and Mystery of
How Humans Navigate the World*

IGNITION

LIGHTING FIRES IN A BURNING WORLD

M. R. O'CONNOR

BOLD TYPE BOOKS
NEW YORK

Copyright © 2023 by M. R. O'Connor

Cover design by Pete Garceau
Cover photograph © iStock/Getty Images
Cover copyright © 2023 by Hachette Book Group, Inc.

Hachette Book Group supports the right to free expression and the value of copyright. The purpose of copyright is to encourage writers and artists to produce the creative works that enrich our culture.

The scanning, uploading, and distribution of this book without permission is a theft of the author's intellectual property. If you would like permission to use material from the book (other than for review purposes), please contact permissions@hbgusa.com. Thank you for your support of the author's rights.

Bold Type Books
30 Irving Place, 10th Floor New York, NY 10003
www.boldtypebooks.org
@BoldTypeBooks

Printed in Canada

First Edition: October 2023

Published by Bold Type Books, an imprint of Hachette Book Group, Inc. Bold Type Books is a co-publishing venture of the Type Media Center and Perseus Books.

The Hachette Speakers Bureau provides a wide range of authors for speaking events. To find out more, go to hachettespeakersbureau.com or email HachetteSpeakers@hbgusa.com.

Bold Type books may be purchased in bulk for business, educational, or promotional use. For more information, please contact your local bookseller or the Hachette Book Group Special Markets Department at special.markets@hbgusa.com.

The publisher is not responsible for websites (or their content) that are not owned by the publisher.

Print book interior design by Linda Mark.

Library of Congress Cataloging-in-Publication Data

Names: O'Connor, M. R., 1982– author.
Title: Ignition : lighting fires in a burning world / M.R. O'Connor.
Description: First edition. | New York : Bold Type Books, 2023. | Includes bibliographical references and index.
Identifiers: LCCN 2023027045 | ISBN 9781645037385 (hardcover) | ISBN 9781645037378 (ebook)
Subjects: LCSH: Wildfires—Climatic factors—North America. | Wildfires—Prevention and control—North America.
Classification: LCC SD421 .O29 2023 | DDC 363.37097—dc23/eng/20230703
LC record available at https://lccn.loc.gov/2023027045

ISBNs: 9781645037385 (hardcover), 9781645037378 (ebook)

MRQ-C

10 9 8 7 6 5 4 3 2 1

For Bryan, the spark

CONTENTS

PROLOGUE		1
1	PRAIRYERTH	13
2	FIRST, LIGHTNING	41
3	PYROPHILIA	63
4	MONSTROUS SERPENTS	87
5	EMISSARIES FROM A DIFFERENT REALM	109
6	DIRTY AUGUST	139
7	EVERYTHING IMAGINABLE HAPPENED	165
8	CIRCLING THE SQUARE	185
9	RESIST, ACCEPT, OR DIRECT	207
10	DRAGON EGGS AND HAMMERSTONES	229
11	BEAUTIFUL AND RIGHT	247
12	IN THE LAND OF THE WHITE DEER	273
13	A PRAYER WE MAKE TOGETHER	293
	EPILOGUE	311
	Acknowledgments	325
	Notes	329
	Bibliography	339
	Index	355

It still holds me fast; and I shall tend it while I can; while I live, I shall love to gaze on this ancient soil.

—Gurra, as told to T. G. H. Strehlow

PROLOGUE

WILDFIRES HAVE BECOME A SYMBOL OF CLIMATE APOCALYPSE, FREQUENTLY CITED as unambiguous evidence of the fossil-fuel induced ecological disaster that is destabilizing of life on earth. Get close enough to a fire on the landscape—maybe even close enough to smell it burning and feel its heat—and what you may find is a far more complicated and fascinating story about evolution, biology, history, and culture. And one that is infinitely more hopeful.

My first encounter with fire was in Australia. I was driving in the Northern Territories on my way to interview an Aboriginal man on his ancestral homeland, and my route took me through a forest whose appearance was so strange to me it seemed almost alien. The trees were eucalyptus, or as Australians call them, gum trees, of different varieties—scarlet, poplar, ghost, river, and white. Each individual tree was spaced so far from the others that their branches never seemed to touch, making the forest look too sparse, maybe even blighted. Strangest yet was the ground underneath the trees. It was barren of grass and black from the scorching heat of a recent fire. It looked awful to me, but when I arrived in the town of Katherine that evening and expressed my concern to a local about the apparent disaster that had taken place, their response surprised me. Oh no, you don't understand, I was told. People lit that fire intentionally. Around here, they love to see the earth blackened. People take pride in lighting fires to keep the land clean. It's a sign that the country has been cared for. I had never heard of people starting fires on purpose.

When I flew back to America a couple of weeks later, I sat on the plane reading a book by the Australian historian Bill Gammage called *The Biggest Estate on Earth*. Gammage documents the complex land-management practices of Australian Aboriginals before European colonization in 1788. Their primary management strategy was igniting fires. "Fire let people select where plants grew. They knew which plants to burn, when, how often and how hot. This demanded not one fire regime but many, differing in timing, intensity and duration," Gammage wrote. "No natural regime could sustain such intricate balances.... With fire as an ally people worked the land as intimately as humans can."

Gammage described a relationship I'd never imagined. To sustain their country, Aboriginals created a system of law and tradition in which each individual was duty bound to care for their country and had a role in doing so. "The cycles of life and season change constantly, and a manager's duty is to shepherd land and creatures safely through these changes," he wrote. "This might require dramatic or spasmodic change (burning forests, culling eels, banning or restricting a food), and it certainly demands active intervention in the landscape." At a continent-wide scale, Australia was a sophisticated mosaic of plants and trees of different ages that had been created by people manipulating fires of various intensities in different seasons. They were pyrotechnicians.

IT WAS A few years after I read Gammage's book that Australia underwent the Black Summer, a season of bushfires that engulfed swaths

Prologue

of the drought-stricken continent in flames. Over the course of a few months in 2019 and 2020, forty-six million acres burned. Ecologists warned that as many as three billion vertebrates had been killed or displaced, some even driven to extinction. Fire scientists were aghast at what they called a "super outbreak" of fire-induced and smoke-infused thunderstorms that were contributing to extreme fire behavior. Over a few months, they documented an unprecedented thirty-eight of these storms. Two of them produced the largest smoke plumes ever observed, so big they broke into the earth's stratosphere and were consistent in magnitude to a moderate-sized volcanic eruption. More than seven hundred million tons of carbon were released during Black Summer, an amount that exceeded the annual carbon footprint of some industrialized countries, including Germany and South Korea.

In the midst of the crisis, I wrote to an acquaintance, Jack Egan, in New South Wales, to see if he and other mutual friends were OK. Egan wrote back that his home had been incinerated six days earlier. "The fire impact in Southern Australia is devastating," he wrote. "Amongst the deaths and damage Cath and I lost our house. We are philosophical about that. We have always been prepared for that living in a tall forest as we do." Speaking to a newspaper reporter, Egan described how he sheltered in a neighbor's concrete basement as his house burned in what he described as a firestorm. He linked the Black Summer fires to climate change. "Many of us, including me, assumed the future would unfold in an orderly way, gradually getting hotter and more uncomfortable and we'd adjust," said Egan. "The future has crashed right into the present now in a chaotic way—it's not orderly in any way. Clearly that's not OK—it's like a war zone down here."

These two moments in my experience—encountering the burned forest outside Katherine and witnessing the scale of Black Summer—became symbolic to me. I sensed they were two parts of the same story. But how? What was the connection between the fires of 1788 described by Gammage—the intricate balance and understanding of fire as a form of stewardship—and the destruction wrought by the Black Summer? I found an answer in the words of Bill Neidjie, a Gagadju elder. Decades before, Neidjie had lived through the Ash Wednesday fires of 1983, which destroyed thousands of homes. When asked what he thought about the fires, he answered that it was a "crime" to allow a country to get so dirty. "If people don't take of their country, if they don't burn it, keep it clean, then it grows and grows, and when it does burn it's a really hot fire. I take care of my country. We don't (have) fires like that."

ONE DAY LONG after I returned from Australia, I walked on the side of a dirt road in Florida's Apalachicola National Forest with a local botanist. We had set out that morning in search of carnivorous plants, but we took a detour, parking the car to explore a grassy ditch. Now we stood next to each other and stared at the electric-blue petals of an extremely rare wildflower, *Gentiana catesbaei*. Trash was strewn about, and cars sped past us, oblivious to the precious jewel growing on this random roadside. "The only reason it's here," the botanist mused, "is because they mowed under the power lines and mimicked a wildfire."

Wait, I thought. So, some wildfires are good? In the land of Smokey Bear, where wildfires are always bad, this line of questioning felt almost verboten. But the flower led to a paradigm shift in my understanding of wildflowers and ecology. *Gentiana catesbaei* is a pyrophyte: it tolerates and even likes fire, which creates the open conditions for the sun-loving plant to grow. Eucalyptus trees like those I saw in the Northern Territories are also pyrophiles. Their oil is highly flammable, and their cones are sealed with resin that can only be melted by intense heat, which releases their seeds for germination. These kinds of evolutionary alliances between plants and fire are hundreds of millions of years old. First, plants made fire possible by providing the fuel for combustion to take place. Then plants adapted and even exploited fire to their benefit.

Certain episodes of our planet's history were especially fiery. The Cretaceous period, for instance, had a warm and highly oxygenated atmosphere and overlapped with the dominance of angiosperms—the flower- and seed-bearing shrubs, grasses, and tree species that we live with today. Angiosperms were rapid photosynthesizers that grew fast and sprouted small seeds that spread and proliferated easily. The aerated fuels of angiosperms were flashy and burned hot and quick, traits that created the conditions for more fire... and more angiosperms.

Ancient fires played a pivotal role in the evolution of our planet's biomes. Around seven to eight million years ago there was an explosion of grass angiosperms across the world. Scientists know this because these plants have a unique photosynthetic pathway, the process by which they convert light energy into growth. Their pathway creates a four-carbon rather than a three-carbon compound. When they analyze sediments, scientists find this carbon signature in the grasses that spread across Asia, Africa, and the Americas. It was in these frequently burning savannahs that the human species evolved.

Fire is extraordinarily complicated. It is difficult to categorize, impossible to simplify. As some wildland firefighters put it, fire is inherently good... up until it isn't. Where that demarcation lies always depends on the context. Earth abounds with pyrophytic species that self-immolate and then flourish. Like eucalyptus, the European *Pinus halepensis* has serotinous cones that can't germinate without fire, even though adult *halepensis* trees are especially sensitive to fire. These fully grown pines are destroyed to make way for the next generation to grow. Similarly, lodgepole pines depend on high-intensity fires that kill many adults but also open and disperse their seeds into the earth to sprout. Shortleaf pine buds lay dormant underground, waiting until a fire has burned, and then emerge quickly using the nutrients stored in the root system of the trees. Snowbrush ceanothus, a type of wild lilac found in North America, has seeds that can remain buried for two hundred years until a wildfire awakens them by scarifying the hard outer layer. Many orchid species flourish following wildfires, proliferating in the mineral-rich ash beds and bathing in newly opened canopies. The 2009 Black Saturday bushfires in Australia stimulated so much growth in threatened orchid species—white caledenia, leopard doubletails, green leeks, and spotted sun orchids—that the flowers blanketed parts of the charred landscape. For two thousand families who lost their homes during the Black Saturday bushfires, fire was devastating. For orchids, Black Saturday created the conditions for new life.

I SET OUT to understand the ecology of fire. What I discovered was a kind of natural history of arson. Intentional fire-setting by people was once a widespread land-management tool, practiced the world over. Australian Aboriginals deliberately set fire to the land, augmenting natural fire regimes and even manipulating them for their purposes. Indigenous North Americans used fire to shape and expand prairies, create orchards of chestnut and oak, and craft quilt-like mosaics of meadow and woodland.

The myth of wilderness is tenacious, but this secret history of fire negates the fiction of a natural world untouched by humans. As the philosopher Martha Nussbaum points out, the wilderness ideal comes from the Romantics, who envisioned a man separating himself from society and renewing his spirit out in the woods through the experience of wonder and awe. "Nature, in this conception," wrote Nussbaum, "is supposed to do something for us; the idea has little to do with what we are supposed to do for Nature and other animals." Nussbaum calls wilderness a narcissistic concept.

By mimicking natural processes for cultural objectives, igniting fires subverts distinctions between culture and nature. Human fire-lighting also recognizes a complex yet basic reality of life on our planet: so much of its health and flourishing depends on cycles of death and renewal. Fire is an agent of those liminal transformations. It disturbs and kills and midwifes the new into being. Acknowledging these contradictions, the anthropologist Donna Haraway has said that fire is an "essential element for ongoing," even as it presents a "double death, the killing of ongoingness."

One of the impacts of the genocide of Native Americans was that it disrupted traditional land-management practices that included widespread fire-lighting. As early as 1634, European settlers on the continent were noticing that the death of Native Americans from disease was changing the now unattended landscape. "In some places where the Indians dyed of the Plague some fourteen years ague, is much underwood, as in the mid-way betwixt Wessaguscus and Plimouth, because it hath not been burned," wrote one settler. As the historian Stephen J. Pyne wrote, "America was not a virgin land but a widowed one, emptied of indigenes by conquest and disease. Wilderness was not an intrinsic characteristic of pre-Columbian America but one fashioned by settlement and industrial America."

For more than a century, the dominant practice in the United States has been to suppress fire in order to protect forests and communities. This approach has intensified the very problem it was trying to solve. Preventing "fuels"—grass, shrubs, and trees—from burning today only preserves them to burn tomorrow. As the stockpile of fuel grows, fires burn longer and with greater intensity. Suppression has created a fire deficit and a need to reduce the fuels that cause high-severity wildfires. In California alone, an estimated twenty million acres—an area the size of Maryland, Massachusetts, and New Jersey combined—would need to burn to eliminate the fire deficit created by a century of suppression. Federal agencies acknowledge the problem, but bureaucratic risk aversion and budget constraints, among other things, have stalled the adoption of new approaches, leaving America both burning and fire starved.

Fire ecologists have a term for the fact that putting out fires contributes to the creation of even bigger blazes: the fire paradox. The megafires we see on the news are examples of the sins of the past haunting the present, sins that predate the combustion engine by hundreds of years. But global warming *is* exacerbating the paradox: climate scientists predict that, by the middle of this century, the number of days conducive to wildfires of enormous size and complexity, the aforementioned megafires, will increase by as much as 50 percent.

As a journalist, I wanted to understand the fire paradox by talking to people whose lives are intertwined with this powerful element in profound and sometimes tragic ways. When did humans start lighting fires? Why did we start putting them out? How much has the erasure of other histories and perspectives created the monstrous conflagrations we see today? (What is it like to fight a monster?) Is it possible that many of our current environmental problems are at their root epistemological problems? How can we escape the paradox?

AFTER THE BLACK Summer, I enrolled in an online course called RX-310, Introduction to Fire Effects, at the University of Idaho. It

Prologue

was designed to teach students the relationships between fire regimes (the frequency of natural fires in an ecosystem) and fire effects (the physical, biological, and ecological impacts on the environments). Over six weeks, we studied things like flora, fauna, plant disease, watersheds, insects, and soils of different ecosystems. We learned that an estimated 80 percent of North American vegetation is fire dependent. While I was otherwise an adequate student, I could not seem to grasp the most fundamental aspects of fire behavior. My professor was a fire ecologist, wildland firefighter, and mother of three who did not hesitate to strap a baby to her chest and demonstrate proper use of a drip torch. I didn't know any wildland firefighters, but she was not the grizzled person I had pictured. In class, the professor would say things like "Increased fuel loading might not increase intensity, but it would probably increase severity and residence time." I would stare into the distance, squinting my eyes, trying to conjure in my mind's eye what this possibly meant. I had no reference points. I had never seen a wildfire or any active fire outside a hearth. It was all a blank. The professor tried to encourage me. "It's OK!" she said. "Your brain is like a dry, crusty sponge. You just need to see it in person!"

I decided to try and see a fire. Maybe I could volunteer on a crew for a day, use a tool, and get a sense of the work. But to get on the so-called fireline, I needed wildland-firefighter qualifications. An endeavor that was initially a bit of a gimmick became something more like a personal odyssey that took over my life: I embarked on what became a thirty-thousand-mile journey into our national back-of-the-beyond to burn in prairies, ponderosa forests, marshes, pine barrens, pocosin, hardwood bottom swamps, and oak woodlands. Many of the people I met in those places are reviving a relationship between humans and fire. This movement of fire practitioners is also happening around the globe and is largely being led by Indigenous communities, who speak of their right as stewards of the land to once again practice "cultural burning." In places as disparate as Australia, Venezuela, Guyana, Canada, and the United States, Indigenous communities are asserting their traditional ownership of lands and setting fires to mitigate wildfire risks and support food-gathering and traditional practices. Fire is an expression and assertion of sovereignty. As Valentin Lopez, chairman of the Amah Mutsun Tribal Band in California, has said, "Fire is a gift from Creator for the stewardship of the land."

It is a remarkably diverse movement. In places like Texas and California, community burn associations are pooling their resources to manage rangeland or heal from the trauma of extreme wildfires. In states such as South Carolina and New Mexico, nonprofits and stakeholder alliances are recruiting and training a new workforce of young fire lighters. From Montana to Maine, national parks are trying to restore natural ecosystem processes that support biodiversity and resilience. Some of the people most dedicated to the practice of fire-lighting live in America's heartland, far from the stereotypical centers of environmental conservation or climate change concern. In 2023, the American states intentionally burning the most acres were Georgia, Alabama, and Florida. There are different names for these kinds of fires. Ecologically beneficial fires. Cultural fire. Controlled burns. Managed wildland fire. Prescribed fire. Green fire. Good fire.

Soon after I finished my wildland firefighting training, I heard that a group of wildland firefighters, land managers, and biologists were gathering in a little town in Nebraska to set good fires on the prairie. At that time of year, the roots, rhizomes, and bulbs of the grasses and forbs are still safely underground. The fire above the roots clears away last year's dead detritus. The flames kill young cedar saplings that encroach and threaten the prairyerth. The ashes and charred organic matter that are left behind nourish and recycle nutrients into the soil, unlocking ammonium, nitrogen, potassium, and phosphorous, a "nutrient pulse" that encourages root growth that stores more carbon. The newly blackened earth absorbs the sun's rainbow wavelengths, warming the soil and signaling to the plants—phlox, blazing star, butterfly weed, meadow rose—that it is time to awaken and drink the light of the big sky.

So, I set out to join them.

1   PRAIRYERTH

THE ROADS THAT LEAD TO THE LOESS HILLS SEEM LIKE THEY WERE DESIGNED BY someone playing with an Etch A Sketch, one perfect right angle after another. At the start of the six-hour journey, the roads followed alongside cornfields and feedlots, and then somewhere past the hundredth meridian, the land turned hummocky. Yellow meadowlarks shot out of the ditches like atoms. The only things that did not follow a straight line were the branches of the Loup River I passed over, named by French trappers after the Skidi or Wolf (*loup* in French) band of the Pawnee.

It was dusk when I reached the town of Ord, nestled in a bend of the North Loup. I pulled up outside a red-brick building on the handsome town square with big shopfront windows, through which I saw people already inside, eating dinner on long tables set in a U shape and facing a wall decked with a whiteboard and maps.

I grabbed a plate and food from the buffet and listened to the chatter as I ate. I noticed with some relief that there were two other women on the crew. The room had an old tin ceiling and was already littered with radios, coffee cups, duct tape, maps, printers, Sharpies—the detritus of fire operations. This was the crew's incident command post, the place where the "overhead," crew leaders, would hold morning briefings and evening debriefings. These so-called after-action reviews involved discussing, often in great detail, the events of the day, scrutinizing them for successes and failures. I felt nervous. There were sixteen people on the crew, and nearly everybody except me was a seasoned wildland firefighter. They were there to help local landowners ignite prescribed burns but also to work on their qualifications to advance in their wildfire careers. This kind of event was a training exchange, or TREX for short.

All I knew about wildland firefighting had come from the core wildland-firefighting classes known as S-130 and S-190. During the forty hours of online work, I was baptized in jargon. Anchor points. Ladder fuels. Mop-up. Scratch line. Drop zone. Snags. Wet line. Swamping. Torching. Sling load. Hotspotting. Backfire. Burnout. Cold trailing. I learned hose lays and nozzle types and how incident command systems work. I was taught the critical watch-out situations known as the "tens and eighteens," the hard and fast rules to prevent casualties on the fireline. Even these seemed obtuse to a neophyte. "Fight fire aggressively, having provided for safety first." The lessons were delivered by an anonymous male narrator who switched between wry sarcasm and scolding. "If you're under the impression that firefighting is a literal walk in the park or a nice job to perform in the great outdoors," he said, "realize now that you'll need to spend as much time training your mind as you will your body."

I took the multiple-choice tests at the end of each module and received certificates of completion. Then I took the required Work Capacity Test required for arduous fireline work: carrying forty-five pounds for three miles in forty-five minutes or less. To do this in Brooklyn, about as far from wildfire country as imaginable, I filled a backpack with sandbags and speed walked around a park with my seven-year-old son jogging beside me.

After dinner I borrowed personal protective equipment, or PPE, from the incident command's cache: a radio, a hardhat, a fire pack containing a fire shelter (a tent made of aluminized cloth, designed to protect occupants from temperatures as high as seventeen hundred degrees), and a yellow coverall made of fire-resistant material called Nomex. This "banana suit" would be my uniform on the fireline.

I stepped outside and started walking to my car when I heard someone shout behind me, "Are you the journalist?"

A woman in her early fifties stood on the sidewalk in a baggy blue sweatshirt and baseball hat, her light brown hair tucked behind her ears, smoking a cigar. Standing next to her was a guy in his sixties wearing a straw fedora and puffing on a cigarette. They cheerfully waved me over.

"I'm going to watch what I say around you!" the woman said. Her name was Patty Carrick. She was from the Upper Peninsula of Michigan, I found out, and had driven fifteen hours to Ord towing a UTV hitched to her truck. The man's name was Dan Kelleher. He had driven to Nebraska from Chico, California. Twenty hours.

As we stood outside in the dark, Dan and Patty reminded me of a pair of teenagers sharing a joke in the high school parking lot. Halfway through the conversation, I realized they were the principals of the whole school, the overheads in charge of burning four thousand acres of prairie over the next two weeks. I didn't know what I had expected, but it wasn't two nicotine addicts cracking jokes into the night. As I listened to their banter, an impression began to form. Dan and Patty seemed so... happy. It startled me, this feeling that they knew or had something I didn't. What was it?

Over the coming days, and in the years that followed, without their permission, I adopted Dan and Patty as my role models. They were irreverent, joyful, and hardworking. They possessed an inexhaustible amount of practical knowledge, the riches of countless days spent in the woods with tools, chainsaws, and drip torches. They were great friends to one another, affectionate, conspiring, and supportive. I once heard Dan call over the radio at the end of a long day, "Carrick. Kelleher."

"Go ahead," replied Patty.

"Have I told you I loved you today?"

Patty's passions were racing vintage snowmobiles and hunting. She had spent her honeymoon rafting sixty miles of a remote river in Alaska with her husband, Joe, an Ojibwa member of the Bay Mills Indian Community, hunting moose during the day and boondocking every night. But Patty really lived for fire—starting fires, putting them out, it didn't matter, so long as she was close to the action. She was the assistant fire management officer for the Hiawatha National Forest, responsible for conducting controlled burns and fighting unwanted fires. Before that, she had been an engine boss leading a crew.

The US Forest Service, the agency under the US Department of Agriculture responsible for managing 20 percent of the nation's forests and the majority of its wildland firefighters, is a male-dominated agency. Patty's boss, it turned out, was one of just a few women in the country who had the title of fire management officer. "I guess it's unusual to have two middle-aged women running a fire program," Patty told me. "My husband says we have 'too many squirrels.' But we did the first aerial ignitions in the Hiawatha! We get more done than anybody."

It drove Patty nuts to miss a fire in the Hiawatha. "I could be on the best wildfire out west, but if Joe calls and says there's a two-acre fire on the Hiawatha, I'd rather be there," she said. "I want them all. I'm selfish." Over the next two weeks, Patty would be one of the task force leaders as well as the safety officer, communications chief, and training specialist for the crew. I asked her how long she had been coming to Nebraska to burn. "Ten years," she said. "This is my spring break!"

Dan was spry and mirthful, with a white mustache and blue eyes. He had been coming to Nebraska for even longer than Patty. "What's your all-time favorite place to burn?" I once asked him. "Here," he said without hesitation, nodding toward the amaranthine prairie surrounding us.

Dan was a burn boss, a senior position requiring many years, if not decades, of experience and at least fifteen other firefighting qualifications in order to command complex prescribed-burn operations. Over his lifetime, he seemed to have had almost every job in fire. He had worked for thirty years as a wildland firefighter, engine operator, and station captain for the Forest Service before moving to Port Townsend, Washington, to work as a structure firefighter— a municipal firefighter who puts out fires in homes and buildings. In 2004, he moved back to Chico to work for a private contractor and then became the head of their training program, teaching over forty-five wildland fire courses and traveling across the country to lead controlled burns.

Dan liked to describe himself as an "old hippie" and "fucking crazy." It was true he had grown up in California in the sixties and had a penchant for quoting Bob Dylan and Joni Mitchell lyrics. He could be zany. (One of the crew's favorite Dan-isms: "I said wiggle waggle, not squiggle waggle!") But Dan's outward freneticism belied an extraordinary intelligence and mental dexterity. Every minute of the day he was tracking dozens of variables—multiday weather forecasts, terrain, fuel types, smoke, predicted fire behavior, personnel, and safety—and weighing, interpreting, studying, evaluating, ruminating, and reevaluating these factors in concert to decide when, where, what kind, and how much fire to ignite. Dan once told me you had to have reverence for fire: it was easy to guess what it would do but hard to be sure. "It's a very dynamic, hard-to-read force in reality. There's just too many factors."

"What would you have done if you hadn't gone into fire?" I asked.

"I probably would have been a CIA analyst."

Someone later described Dan to me as one of the true and great mentors of prescribed fire of our time. He was a kind of fire savant. Watching him work reminded me of the words of an Aboriginal fire lighter. "You sing the country before you burn it," he had said. "In your mind you see the fire, you know where it is going, and you know where it will stop. Only then do you light the fire."

When it came to Dan and Patty, my thinking was simple: If I spent enough time doing what they did, I might find my way to the same basic goodness they exuded. I might find out what was making them so happy.

ON OUR FIRST morning together as a crew, Dan stood in front of a large whiteboard at the incident command post. "You need to know who you're working for and what your mission is," he said. "You need to know that at all times."

Next to him was the crew's burn boss trainee. (In wildland firefighting a trainee is someone who has all the required experience and training for a position but is still completing an assignment "task book" before they are fully qualified.) The trainee was a quiet, reserved guy in his thirties. He had a wavy blond beard that reached all the way to his chest, and an ever-present bulge of dip tucked into his lower lip. He worked for the US Bureau of Land Management out in Wyoming, but before that he had been a hotshot in Utah for seven years. Hotshots are members of elite firefighting crews dispatched to complex and aggressive wildfires. They often work with basic hand tools to dig "handline," removing trees, roots, and other fuels from a continuous strip of land. Beyond the handline, they ignite their own fires with drip torches, creating areas of blackened ground to starve the main fire of fuel. This process, known as backfiring, is one of the core strategies of wildfire suppression. The trainee rarely talked about being a hotshot. The only indication he had been one was his belt buckle, which showed his crew's insignia, and the unflappable calm he exhibited around fire.

Now Dan and the trainee led a refresher in controlled-burning tactics. The trainee drew a box on the board in red marker. It represented a hypothetical unit to be burned. Then he drew an arrow representing a south wind. "If we have a south wind, where are we going to start our fire?" he asked. He drew a star in the middle of the northern edge of the unit. This seemed counterintuitive to me. If the wind was coming from the south, wouldn't it be easier to start a fire at the southern edge so that the wind could push it through the unit? This was exactly wrong.

Wildland firefighting categorizes fire into three types: head, flanking, and backing. Head fires move with the direction of the wind. They burn hot, generating the biggest flames because the heat of the head fire preheats the fuels in its path. The combustion of head fires generates intense energy, so they have the highest potential for fierce and unpredictable behavior. Head fires also move fast: if you start one, you better have a plan for how to stop it. Backing fires are usually the most predictable. They burn against the wind, and because of this they move slowly. They can also have the most intense effects because of their slow speed: they spend more time in one place, what is called the residence time. Flanking fires are flames that move perpendicular to the wind and, depending on fuels and topography, can have moderate to intense levels of fire behavior and effects.

The trainee explained how, by setting different types of fire at different times and locations, we would direct where the flames would go, how fast, and how intensely. He drew another diagram to illustrate: two "ignitions" teams moving in opposite directions, each laying down strips of backing fire. A "holding" team followed behind, using hand tools and engines to ensure the fire didn't escape the perimeter of the unit. Once the ignitions teams had created enough burned fuel, or what was called the "black," they could ignite a head or flanking fire into it. The trainee called it making a "catcher's mitt" for fire.

"We want our black to be deep enough that, if we get a head fire, it will die off in the black," added Dan. "I don't know if you noticed but there's only sixteen of us."

Controlled burning exploits wind, terrain, and fire behavior so that the fire goes where you want and stops where you want. But no one can control weather. Wind had created the Loess Hills and all its vagaries, the vertical bluffs and gullies. In Nebraska in spring, gusts of forty miles per hour were not unusual; it was one of the windiest places in America. For this reason, the possibility of escaped fires—no matter how careful we were—was real. "Where it's flat, they grow crops. Where it's bumpy, they grow cows," said Dan. "And, when it's bumpy, the wind starts doing strange things." A wind-direction shift could transform a backing fire into a head fire tearing through the grass. "Wind shifts are a fucking big deal," Dan warned us.

Enormous resources were put into making sure we could attack a fire if it escaped a unit. We had a Type 3 engine that weighed twenty-six thousand pounds, had four-wheel drive, and could pump 150 gallons per minute; multiple Type 6 engines (basically big pickup trucks) outfitted with water tanks and pumps; and UTVs with smaller tanks and pumps slipped on. We had bladder bags, collapsible backpacks with portable sprayers, and countless hand tools. The most important of these was the flapper, a long wooden handle attached to a piece of square rubber used for swatting and smothering flames in the grass. "We want to catch these things small because they will run and run," said Dan.

Attacking a ten-foot "slopover" that had escaped a perimeter was no different—in theory—from attacking a ten-thousand-acre wildfire. "Are we attacking the head?" the trainee asked us. "No, we're going for the flanks and extinguishing them first so we're not fighting huge fire. We can't fight ten-foot flame lengths." We would work from an anchor point, the place where the fire had already burned fuels: the black. Assuming you had enough of it around you, the black was the safest place to be. "Sometimes the flanks blow [up], and then we have a new head," continued the trainee. "It gets a lot more complicated when you talk about fuels and terrain. But no matter how big the fire is, you start putting in lines of black wherever you can get into."

In all of our fire packs, we carried fusees, basically dynamite-sized matches that we could strike and use as a torch if we got stuck in unburned fuel (the "green") and the wind shifted. Without any black around, we would have to create our own. "Here's the thing about fusees," Dan said. "If you get it in your eye, you don't have an eye. It's not like it hurts a few days."

We carried fire shelters all the time in the event of a worst-case scenario: getting trapped by an approaching flaming front that we couldn't outrun. I had always imagined forest fires were the most dangerous, but so-called flashy fuels like grass can reach a thousand degrees Celsius, and when whipped by the wind, grass fires don't run, they fly. "Just like at Mann Gulch, that grass burns fast," Dan said. He was referencing the 1949 wildfire in Montana that killed twelve smokejumpers—wildland firefighters who access remote fires by parachuting out of planes. One of the few people to survive Mann Gulch was a guy by the name of Wag Dodge. As the fire raced up the slope toward the crew, he stopped and lit a match and set fire to the grass around him, creating a burned-out area that he lay down in. It saved his life. "Build your own safety zone. If where you are standing is going to get burned, build yourself a safety zone," said Dan. "It sounds crazy but it works." I heard a smokejumper call the strategy "Wag-Dodging it."

I IGNITED MY first fire that afternoon, standing on the slope of a giant dune that had been made by glacial outwash and then sculpted by erosion and time. Why had I always thought of the prairie as flat? It seemed more like an ocean full of whales whose backs I had to scramble to climb up and over. At the top of one hump all I could see were more leviathans covered in golden-cured grass. I shivered in the cold. Everything on the prairie felt raw and exposed. There were no hiding places, only the elements: sun, earth, wind. And now, fire.

"Have you ever used a drip torch before?" asked Dan. Smoking a Marlboro Light in one hand, he held out a battered drip torch, its cherry paint scraped and peeled away.

"No," I said. "But I want to."

"Here you go!" he said. I took it from him.

The drip torch, a metal canister with a handle and a long nozzle, had been filled to the top with fuel and weighed about sixteen pounds, heavier than I'd imagined.

Dan instructed me to pour some fuel out of the torch's nozzle onto the grass. He bent over and lit the grass with a lighter. I dipped the nozzle back into the flames so that its tip caught on fire and then walked a few yards away to a patch of big bluestem whose culms were as high as my waist.

Dan's goal that day was to burn thirty acres of prairie before it rained. There was already a blanket of swollen, grey clouds overhead. He kept looking up to the sky, trying to divine how much time we had before the heavens opened. Normally, he explained, bluestem burns hotter and "flashier" than other grasses, especially when it is dead. In the spring, bluestem leaves and stalks lack the living tissue that holds moisture; in dry conditions, lighting a field of dead bluestem is like setting off a pyrotechnics show, the firecrackers all whistling and exploding in the heat. "It's fun," Dan said. But in humid conditions, dead fuel will rapidly absorb moisture and then stubbornly smolder or not burn at all.

Our window for getting a fire going was closing as the rain approached. "I got missions I want to accomplish!" urged Dan. I pushed the torch's nozzle into the thatch, tipping it to pour out fuel and flame, and then walked a few feet away to the next patch of bluestem. I focused on picking up the biggest patches of grass—"jackpots" of fuel. My heart skipped around in my chest, whether from exertion, pleasure, or fear I didn't know.

When I reached the far end of the field, I looked behind me. The big bluestem patches I had just lit sizzled and puffed smoke; the flames had shrunk and been swallowed up. I felt raindrops.

"I don't think it's working!" I shouted to Dan.

"I guess we gave you the most dangerous job!" he yelled, and laughed.

My first fire did not start. Later, I wondered if Dan had let me fail on purpose. He knew that every fire had something to teach. He urged me to watch fire closely, to "play fire scientist." What matrix of variables created the fire's character? How fast, timidly, unpredictably, powerfully, eagerly, or gently did it behave and why? He taught me to study—from one moment to the next—the interactions of wind, humidity, topography, fuel, clouds, and air and try to understand the myriad ways they conspired to spur or inhibit combustion.

That spring day, I saw for myself that fire is the product of an alignment. In my online firefighting courses, this alignment was called the fire triangle, and it comprises three ingredients: fuel, oxygen, and ignition. When all three are present, a chemical reaction is unleashed, and energy in the form of heat and light is released. Take away just one of the conditions and fire is impossible. I understood that the variables can exist in one instant and disappear the next.

I never forgot my first fire. Before then, I had only really understood fire as dangerous and threatening. Afterward, I knew there were different kinds of fire. Fire as luck. Fire as miracle. Fire as gift.

DAN CALLED FIRE the "earth's cleanser," and he always reminded the crew of our purpose. "Why are we burning?" he would ask. "We apply fire to achieve ecological benefits."

To understand these benefits, we had to go back in time to when prairie fires were common. White settlers described those events in terrifying language. At first, a glow would appear on the horizon as far as forty miles away. Then the wind delivered a blizzard of ash and black smoke that would blot out the sun, driving the farm animals crazy. Finally, a flaming front of flames thirty feet high would arrive at the speed of a galloping horse, so fast there was no hope in trying to outrun it. Eventually, prairie settlers dedicated themselves to putting out those fires. But grass desires fire. Combustion speeds up the decomposition of organic matter and results in the fertile humus known as prairyerth, which then midwifes more grass that attracts wildlife. In addition to putting out fires, settlers also planted trees. Before them, the vast prairies west of the Mississippi River and east of the Rocky Mountains contained few if any trees. They were savannahs, and regular fires killed the saplings that would otherwise transform them into woodland. The practices of the Indigenous inhabitants inhibited tree growth as well: the Pawnee farmed the flat land—growing pumpkins, corn, and beans—and burned the hills so the buffalo would be attracted to the young, green shoots that proliferated.

"The system nowadays isn't that different," an ecologist explained to the crew one day. We sat around the folding tables as rain fell outside the window; it was too wet to burn. "We're still farming the valleys and grazing the hills." But the prairie today is almost unrecognizable from what it was a century ago, namely because there are so many trees now. At first, trees were planted for fuel, shade, and building material, part of the civilizing of the Great Plains, as important as tilling the soil with a plow. Trees became so popular among Nebraskans that the state governor founded Arbor Day in 1872. The largest man-made forest in the Western Hemisphere is in Nebraska. Then the Dust Bowl set off yet another race to plant trees, and soon cedars were subsidized by the government, given to ranchers to use as windbreaks for cattle. Now the cedars were overrunning the grassland, proliferating in pasture, and encroaching on more than a quarter million acres of prairie each year, transforming it into a homogenized woodland. "We almost have this fetish about planting trees," said the ecologist. "We did this to ourselves."

I had never encountered anyone with more animosity toward trees. The ecologist's arboreal prejudice was really against *Juniperus virginiana*, the eastern redcedar. Most of his career has been devoted to trying to destroy them, and he infected us with a similar zeal for their eradication. "You should feel good about killing little baby trees," he said. "So many windbreaks never had a hoof in 'em." Cedars are actually native to the prairie. The town graveyard we passed each day to drive to the crew's motel contained beautiful specimens, strong and tall with lovely old and tough crowns. The stunted, rust-colored cedars we saw as we scouted our burn units were different: they had been genetically modified to grow fast, and once they reached maturity they became virtually fireproof. Our task was to get the fires burning hot enough to destroy the young saplings hiding in the grass.

Ben Wheeler, a wildlife biologist with the nonprofit Pheasants Forever and the organizer of the annual TREX, told me he once counted the little purple berries of a female cedar out of curiosity. He finally stopped at 465 berries—on a single eleven-inch branch taken from a 25-foot tree. Ben was tall and lanky with short brown hair and dark eyes. He had a serious, businesslike disposition that I soon discovered he switched on and off depending on whether he was at work or at the bar. Ben described himself as a "prairie fairy." He had grown up in Indiana, attended graduate school in Kansas, and then started a family in Ord. He could perfectly mimic the sound of a prairie chicken and loved to pause and call to them in the early morning.

Ben had a vision. Bring fire back to Nebraska's landscape at scale, thousands of acres at a time. "At one point, fire just got taken out," he said. "In my area, there has not been a continuation of fire—purposeful, intentional fire—in generations." He believed that the only way to realize this vision was to give drip torches to local landowners. Ninety-seven percent of Nebraska's land is privately owned, so without their effort not much would ever get done. It could be a tricky message to deliver to folks. To do a prescribed burn that would burn hot enough to kill redcedars, landowners needed to rest their grass for a year, grass that cows could be getting fat on. "To not graze is a pretty big economic hit," he explained. "It's not easy to convince people to burn up grass rather than eat it." Ben's job involved starting a conversation with landowners, one that could last months and years. He talked to them about the long-term financial incentives of burning: to improve the forage for their cattle and prevent trees from overtaking their grazing land. He was, in a sense, urging them to adopt the logic of the Pawnee. So far, there were around ten prescribed-burn associations in Nebraska, forming a neighbor-to-neighbor model for sharing equipment, knowledge, and labor for prescribed fire. Most of the landowners who burned eschewed Nomex and government qualifications for denim jeans, common sense, and know-how. Some welded their own trailers out of old truck beds to haul pumps around, their farm dogs yapping on the fireline as they worked.

Ben spent part of his year writing up burn plans and securing permits to create what he called a "fire playground" for the wildland firefighters who showed up each spring. But his real hope was that one day the event wouldn't need to take place. "We have a lot of federally qualified firefighters who wanted to get experience and get qualifications that advance their careers. If there are people who need training all over the world, why don't they come here? We need the burning," said Ben. "But my mind is on how we can turn this back to our landowners. We're just here to provide an example and a model that hopefully they can grow into. I want us to gradually become invisible." Ben estimated that over the course of ten years, the yearly crews he assembled in Ord had treated thirty thousand acres with fire. But he estimated that central Nebraska needed to burn at least fifty thousand acres each year to preserve the prairie that is left, and much more than that to restore prairie that has already been lost. Absent of fire, the landscape would become a sterile and unproductive place. There were two possible approaches to preventing this future: burn often or burn big. As Ben explained, you could conduct 500 one-hundred acre burns or, you could conduct fifty one-thousand acre burns. "I have a dream of road-to-road fires one day," he said. "Our goal here is to dream big."

WE EMPTIED WATER tanks and hoses every night to prevent them from freezing, then set out every morning before dawn to line up the engines by the single fire hydrant to fill them up again. If the pumps were frozen, we would thaw them by boiling water in Jetboils to melt the ice that had collected, or by warming the metal casing of the pump with a blowtorch. We caravanned to an old barn, where I would jump into a UTV, my hands aching as I gripped the cold steering wheel, and follow a simple two-track or mow line across hundreds of acres of prairie. We traversed the leks of prairie chickens, whose short-short-long calls reverberated through the air, an orchestra of avian theremins in the morning light. Then we would wait for the sun's rays to give off enough heat to thaw the frosted grass and for the night's high relative humidity—the amount of water vapor in the air—to drop.

We often huddled in the idling trucks with the heat on. I spent most of my time in the cab of a Type 6 engine with two Forest Service employees. One was tall and thin with floppy brown hair and a laconic air I would come to associate with many wildland firefighters. Like many previous generations of his ilk, he worked all summer into the fall doing intense, physical labor in places like Alaska and California to pay for skiing, surfing, or far-flung travel in the winter months. He liked Phish and to blast the Marshall Tucker Band's "Fire on the Mountain" in the early morning as we drove out to a unit. "If I'm going to pick a fight at a bar, it's going to be to this song," he said.

The other wildland firefighter was thick as an ox. He was a military veteran who grew up in California's ranch country playing football and barrel riding. He was loudmouthed and hysterical, though his jokes were sometimes barbed with the multitude of perceived injustices he had suffered. He was particularly fond of reminding us that he had been suspended from Facebook—repeatedly—for his right-wing views. The physical risk and discipline of fighting fires seemed to have offered him a second home after combat.

Despite their temperamental differences, the two of them formed an instant partnership, running the engine together for sixteen hours a day, sharing battle stories and inane opinions, and taking every opportunity to give me an education in the mores of firefighting.

"Don't *ever* show up to a fire with a clean yellow," said the tall one, referring to the standard yellow Nomex shirts everyone wore. I smirked in the back of the cab.

"No, I'm serious," he said. "Rub that shit in a fireplace. Do whatever you have to do."

It was true, the other one concurred, I should *never* show up with a clean yellow. Another thing they agreed on: I should never, ever attach the carabiner holding my gloves directly to my pants if I didn't want to look like an asshole. The carabiner should always be connected to my fire pack. Also, I should under no circumstances keep my goggles, meant for protecting my eyes when it was too dark to wear sunglasses, on my helmet.

I got the feeling that I already had so many strikes against me as a journalist, so steep a mountain to climb if I were to gain any credibility on the crew, that it was imperative I adhere to these rules. But I also gathered that it was possible to reach a level of authenticity so unassailable that you simply transcended all the rules. One morning, through the truck's windshield, I saw Patty walk by with her gloves attached to a carabiner on her pants.

"Wait, what about Patty?!" I exclaimed to my cabmates.

"Oh, Patty can do whatever she wants," said the tall one.

Eventually, a favorable weather report would come in over the radios, and we would rush to assemble the drip torches, put on our packs, start pumps, grab hand tools. Someone, most often Dan, who delighted in catching us off guard as we tried to organize ourselves, would ignite a test fire, pouring fuel on the ground and lighting it with the "tweaker torch" he had picked up at Pump and Pantry, the local gas station and convenience store. The test fire served an operational purpose: to observe whether the fire's characteristics met management objectives and to verify the predicted smoke dispersion. We stood and watched it burn for a few minutes. Was it burning hot or cool, fast or slow, consuming fuels or leaving patchy spots? Was the wind finicky or staying true? Then we got to work.

Our early days were spent blacklining: creating strips of burned fuel around the perimeters of the burn unit that acted as fuel breaks and therefore barriers to the fire escaping. Everyone, even Dan, agreed that blacklining was tedious. I often felt like I was working on a road crew. Someone would hold a hose from the back of an engine and, walking slowly, spray water on the upwind edge of the perimeter. They were followed by an ignitor, who laid down a strip of fire alongside the wet line. Behind them were the people holding flappers, ensuring that the fire didn't cross the wet line. Meanwhile, a second ignitor followed and laid down a strip of fire a few feet away from the first ignitor's fire, increasing the width of the strip of burned fuel. On the downwind edge of the blackline, one or several people with tools and flappers extinguished the flames of the backing fire once the desired depth, anywhere from ten to thirty feet, was achieved. The work was loud, with the relentless thumping of pump engines and growling UTVs, wind, and radios. It could involve sucking smoke for hours and stomping and breaking apart countless smoldering cow-shit patties. "Caca de vaca!" Patty loved to shout as we worked. It could take a day for eight people to burn lines that added up to a mere twenty to thirty acres.

I actually liked blacklining. A lot. To manipulate fire so precisely, even in gusting winds and across uneven terrain along perimeters that twisted and turned, required incredibly technical thinking about fire behavior. "Every time you turn on a blackline, the vector of wind changes. What was backing fire becomes a head fire," explained Dan. "We learn and we adjust, we learn and we adjust and we get better." We had to decide from one moment to the next how aggressively or slowly to apply fire on the ground and to stay alert for sudden changes in flame length and intensity. All of my senses were awakened, attuned to shifting smoke or anticipating sneaking flames. Each strip of blackline, it occurred to me, was a microcosm for the principles of controlled burning that would apply later, when we burned four or five hundred acres at a time.

The intense aesthetic pleasure of fire, the rich, ashened earth it left behind, surprised me. Our work resulted in undulating squiggles of blackened strips across the hills, a Robert Motherwell painting writ across the land. Wildland firefighters seemed to me macho, hardened, and brutishly strong. It struck me as odd then to see how extremely sensitive they were to what they considered a "pretty" burn. If the fire wasn't burning evenly through the grass and left behind patches of unburned materials, they called it "dirty" and went back again and again to clean it up, like compulsive birds of paradise tidying their courtship areas. Uneven lines were undesirable. Keeping the edges straight and clean seemed to satisfy some childlike mission to color within the lines. Eventually I realized that these aesthetic preferences were not whimsy; anything less than a solid blackline of consumed fuel could provide the fire with an escape route to slither its way through unburned grass and outside the perimeter.

Igniting fires with the drip torch was thrilling, but I liked, just as much, holding the line, ensuring that the flames stayed within our intended perimeters. I could stand in the flames and glowing embers of grass, warming my feet, the incense of diesel fuel and burning grass washing over me. The smell of the prairie on fire was familiar and comforting even though, paradoxically, I had never smelled it before—a mixture of sage and resin, citronella and my grandpa, who taught me how to build campfires. I would watch how the burning thatch gave a final pulse of intense red light just before the embers glowered and extinguished, like little stars exploding before they died. What was the word for that instant? I wondered.

One day, the burn boss trainee pointed to a hillside about half an acre in size with a little muddy pond at the bottom. He instructed me to burn it, starting on an upslope corner and working my way across and downward. Then, to my delight, he just walked away and left me to get it done. I began making little arcs of fire, building up the black at the top of the slope and then crossing from one side of the hill to the other, taking fire with me, back and forth. As I worked, I passed the cranny of a coyote den in the hillside. It was rhythmic work, each arc of flame feeding into another, like fish scales. At the bottom, I stood back and watched the whole hillside burn. I found the prints of a coyote in the soft mud near the pond, and I recalled reading how coyotes and foxes would lope ahead of a prairie fire, catching small rodents as they escaped the flames. Overhead, the sky was a brilliant blue; where it met the earth, all was flame and smoke. Everything in view was moving. It was dazzling.

DAN ESTIMATED HE missed just two or three sunrises a year. "I like to watch the dark slip to light," he said. "That's my time."

In Ord, he would light a cigarette in the dark and sit at the picnic table outside the motel and feel for the wind. He might cross the street and rough the grass with his hand. "There's a little frost down at the stumps but the stems are good," he would declare. I watched him look at the day's weather reports and compare his own observations, like a high priest of Ra trying to foresee the future. He liked to say he had a crystal ball but it was murky and full of cracks. He seemed always to be negotiating between facts and data and his own intuition. Sometimes, he drove out to a unit in the moonlight, problem-solving difficult terrain or a wind-direction forecast until insight struck. "That's what fire is all about. We've got all this science, and I believe in science, but I don't think we know enough," he told me. "Sometimes the science is saying it won't burn, and it's burning just fine."

Everyone wanted to burn big every day, if possible. It was up to Dan to decide whether a burn was a go or a no-go, which he did by using a checklist of factors. Dan, for all his brash talk and eagerness, was not "torch happy," a derogatory term for someone who put too much fire down too fast.

"How do you make the final decision if something is a go/no-go?" I asked him.

"This, right here," he said, pointing to his gut.

At eight a.m. one day, the crew drove to a unit the eastern edge of which we had blacklined the day before; there was still about eight hundred acres to burn. The sky was a sheet of silver cirrostratus clouds that turned the sun into a watery orb, pale and wan. The night's cold air had pooled in the valleys and created a fog that wound through the frosted hilltops. The black, serpentine lines we had created were so dark it was like staring into negative space, a void we had opened in the earth.

We started on the southern edge of the unit, and I followed a barbed fence with the drip torch, laying down a strip of fire. The grass was already greening up in places, which created a dirty burn, so the burn boss trainee instructed me to go deeper into the unit, creating a bigger head fire that might burn hotter and consume more fuel. The farther away I walked from everyone, the stranger and more elated I felt. The only sounds I heard were the crackling of my radio. I darted in and out of small gullies, searching for the jackpots of fuel, until I reached the southeast corner. A second ignitor and I began working our way in concert up the eastern edge of the unit, lighting flanking and then big head fires. The other task force was setting fires at the far western edge of the unit, and when I paused to survey my surroundings, I had a feeling of being encircled 360 degrees by flames.

I thought a lot about a book I had found called *Ceremonies of the Pawnee*, written in 1921 by James Rolfe Murie, also known as Sa-Ku-Ru-Ta. Murie was the son of a Skiri Pawnee mother and a white father, and as an adult he conducted extensive ethnographic work among the Pawnee. In the book Murie described how the Pawnee believe that in winter, all things are dead or asleep, their forms lifeless. Come spring, the power of life arrives in the form of thunder and lightning, awakening the empty forms and revitalizing them. "Then as the things of the earth awaken and develop, they are cared for and clothed by the winds and the clouds," Murie wrote. This spring awakening reenacts the journey that the Wonderful Being undertook at the beginning of the world. Murie transcribed a ceremonial song associated with the journey:

> *The earth, the earth that is coming.*
> *Wonderful the earth that is coming.*

By midday, the other task force's backing fires and our fires were moving closer together in the interior of the unit. I stood on the rim of a basin full of billowing smoke, like a volcanic crater. Flames lapped at my feet. I aimed a shiny silver pistol over the edge, shouted, "Firing," pulled the trigger, and felt the recoil snap through my hands. As the crack of the shot echoed, an incendiary projectile trailed sparks through the air. It landed somewhere in the void below me, igniting a new fire. We were transforming the prairie into waves of combustion as far as the eye could see.

At dinner, Dan said, "I may take the whole damn unit tomorrow. So be ready for that! I have a deep desire to get my feet back in that ground, touch the earth, do some good out there."

Ben often instructed us not to go back to the motel first if we were going out for beers. He wanted us to drink smeared with ash and dirt so we could be emissaries of good fire in the town. "Don't take a shower, don't change into clean clothes," he said. "I expect you to go smelling like smoke and tell a good story to the locals."

"Stink and drink!" shouted someone. "I love a stink and drink."

Back at the motel, I noticed that my body was different. Now I was the stalk of grass being cured on the prairie. The sun and cold air had turned any exposed skin chapped and leathery. The wind was tilling crinkles around my eyes into furrows. My hands seemed aged by decades. They were my mom's hands.

EVEN DURING A good fire, the stakes felt high. Fatigue, forgetfulness, lapses in judgment all had outsized consequences they wouldn't have in my daily life. Any of them could be catastrophic. As the days went on, the morning briefings seemed to include more and more safety messages. Stay hydrated. Get enough calories. Check your radio batteries, fuel, hoses, PPE. "It's going to be a very dynamic day," said Zeke Lunder, the TREX's planning operations chief and a prescribed-fire activist from California, one morning. "I know Dan enough to know we're going to light some shit on fire. Be ready to roll. Our objectives have not changed—number one is staying safe. It's day eight. Let's take care of each other. Keep in mind that the person you may be running with may not be working at full mental power. If you start feeling shitty in smoke, ask someone to trade. A major objective is minimizing smoke exposure."

"We're going to be working in light, flashy fuels," added Dan. "Our escape routes and safety zones need to be very quick—we're looking at fifteen seconds."

"You should never be surprised when shit blows up," said Zeke. "We're in that ninety-ninth-percentile situation where explosive fire behavior is a possibility."

"You are your own lookout. Keep your eyes up," added Patty. "If you see something, say something. It doesn't matter your experience or lack of experience. If you feel something's not right, say something."

Dan liked to say it was the little things that, if you ignored them, would bite your ass. What I gathered from this was that small things are just stupid mistakes until they accumulate; then each of those small things was a big fat omen that you had ignored at your own peril.

The problem was, I didn't know how to tell a mistake from an omen. Every day, running from an engine to a UTV to the fireline, I seemed to misplace some essential piece of my PPE and spend crucial moments locating it: a glove, sunglasses, my hardhat. I interrogated each lapse: was it thickheaded or a warning? Patty's rule was three bad things had to happen before they added up to a bad sign. But Patty had a knack for spinning bad things into good things, so she never ever got to three. Her pump would freeze, and she would say, "That's OK, we fixed it! That didn't really count!"

I had less tolerance than the others for intense heat and smoke. Not knowing about carbon monoxide exposure from the smoke, I couldn't figure out at first why I woke in the middle of the night with debilitating headaches. Sometimes I had to fight a primal instinct to run from the fire. My brain was like a startled animal, always on the edge of flight or fight, unable to distinguish between exhilaration and terror. One day I found myself holding line when the wind began to pick up, shifting and churning into little whorls. A young wildland firefighter from Oregon and I began aggressively trying to extinguish the flames that now pushed past us and sent clouds of thick smoke up into our faces. For a moment, I thought we were about to lose our line. Dan saw what was happening and switched the direction of our operations so the flame lengths were tempered, and the smoke blew away from us. "That was intense," I said. The young firefighter looked at me, smiling almost sheepishly. "I kind of like it when it gets like that." I had no idea what he was talking about. But when I eventually did feel the thrill of attacking fire, choking on smoke and the heat on my face, I always thought of him.

Later, I wondered whether fire's potential dangers were part of its allure. When I first got interested in training as a wildland firefighter, I had been in the midst of a depression. It had begun with a miscarriage, and then my sadness transmogrified into something darker. The thought of nonexistence slipped into my thoughts with alarming ease and randomness. I became uncharacteristically preoccupied with fears of losing my other children and my partner. Then a friend committed suicide by hanging himself from a tree. (What kind? I was strangely desperate to know.) Among other problems, he had recently undergone intense stress when a wildfire that was part of the August Complex—California's first gigafire, over a million acres in size—came so close to his house that it burned down his garage. His death contributed to my sense, the spring I arrived in Ord, that I was being swallowed up by grief.

Identifying actual dangers on the fireline was, surprisingly, a relief. It felt good to confront risk directly rather than try to make sense of wild anxieties my serotonin-starved mind conjured. Every decision I made for my safety on a fire was life-affirming.

Fire was revelatory in other ways. I still thought about death a lot but from another perspective—how it can be a necessary condition for rebirth. It was an ecological principle that seemed to work on me like a spiritual balm.

I WAS FAR too inexperienced to have any gut instinct or an omniscience about fire. But on one occasion I had a sense of such foreboding that it was the only time in Nebraska I came close to real fear. I was with Zeke and three other firefighters. We rode in a truck down into the interior of a burn unit to a windmill at the mouth of a gulch between two hills. The terrain was too rough to be lit by an incendiary or UTV. It had to be done on foot. Dan suggested we walk into the drainage, bringing fire with us to light the slopes. "Then," he said, "get back to the truck and drive out quickly."

We parked in front of the stock tank and turned off the engine. We were far away from the pumps or engines, so it was oddly quiet. The only sound was the creaking windmill turning in the breeze. We took torches from the truck bed and paused to size up the terrain. Before us was what looked more like a small box canyon with steep slopes that connected in the back; the only egress was back through its mouth. Even I knew from my elementary wildland-firefighting courses that fire can explode in a canyon, erupting as the heated air concentrates and flames move up unburned fuel on its slopes. I listened as everyone began offering their opinion. Was it best to bring fire *with* us as we walked in? Upslope winds burn fast and hot. The wind direction seemed more terrain-dominated down here. If it shifted, would the flames lie over us as we tried to get back to the truck?

"I'm just thinking about this old firefighter that I knew," said Zeke. "He said something about how you should never bring fire with you into a canyon."

They came up with a new plan. First, we would walk to the back of the drainage and then, on our way out, light fires at the bottom of the slopes, get in the truck, and drive out quickly.

It worked, perfectly. But I remembered the sound of the creaky windmill for a long time, how spooky it was to casually converse about choices that had potentially dire outcomes. Or had it not been dangerous at all? Were my perceptions warped by inexperience?

Zeke and I talked about Marin County, California, where I'd lived as a kid. He had recently taken a bike ride down the coast through neighborhoods I had roamed freely on skateboards with my friends. Zeke remembered feeling that the whole county—crammed with houses surrounded by trees that stretched continuously up into the hills—was doomed to burn in a wildfire soon. "Eventually a north or east wind event is going to take it all out," he said. I wondered if it was a special type of curse afflicting wildland firefighters to be unable look at a landscape and unsee how it would burn.

Zeke was a member of the prescribed-burn association in Butte County, California. Before that, for two decades, he had created wildfire cartography tools for suppression operations. He told me how in 2015, California's Valley Fire engulfed 76,000 acres and killed four people. Zeke had worked to exhaustion for weeks, once on a shift that lasted thirty hours. "When the fires are that big, they don't fit on standard maps," he said. There was one image, of a newly homeless family on the side of the road with a newborn baby, that he could never fully shake from his memory.

A few years later, California's Camp Fire burned more than 150,000 acres, destroyed some twenty thousand structures, and killed more than eighty people. The fire reached within a couple of miles of Zeke's house, bringing with it mass trauma. A friend who had helped rebuild his community following a wildfire in 2008 died by suicide after watching it burn again. Zeke himself experienced everything from rage to despair to depression. "This town that I knew pretty well was just gone," he said. "It was like Dresden or Nagasaki. Just chimneys and rubble and people looking for bodies." Zeke developed post-traumatic stress syndrome from these experiences, including severe anxiety attacks that persisted for years.

From talking with Zeke and others, I began to deduce how complex working with fire could be. In Nebraska, everyone believed in fire's ecological benefits and was committed to its reintroduction into the landscape. "Start thinking about how you apply this work to other places in the world—when you go home to your land, to your community—by helping someone out," urged Dan. But this wasn't a feel-good exercise or what one smokejumper sarcastically described to me as "hippie shit." Some people were trying to save their homes and communities by creating a new fire culture. They were trying to forge an alternative future.

ON MY LAST night in Ord, we sat in the parking lot on the fenders of the engines, drinking under a full moon. In the morning, I caravanned with the crew out to the prairie for morning briefing. Then everyone drove away to start burning. I was envious and waved lamely as they disappeared down the road. I drove in the opposite direction, passing hundreds of acres of hills that we ourselves had painted black, an inversion of the world now illuminated by the rising sun. To the west, the moon still hung over the landscape. To me, it looked right.

2 FIRST, LIGHTNING

I HAD THOUGHT OF FIRE AS A THING, A NOUN, SOMETHING SOLID THAT EXISTS IN space. *The* fire. Seeing fire on the landscape dispelled this idea. If I looked at the flames long enough, the fire began to dissolve, to lose its boundaries and become nebulous. Combustion was not a thing but a moment. A flash of energy transfer, an atomic-level chemical reaction of fuel and oxygen releasing heat and light. Then it disappeared. A wildfire, I realized, was an assemblage of these instants, not a thing but a process happening in time, and no two were the same.

I found an illustrated children's book called *The Book of Fire*, in which the author shows how there are analogues for fire's basic chemical reaction everywhere. A star's fusion of hydrogen and release of energy. A plant's absorption of sunlight and atmospheric carbon dioxide and exhalation of oxygen. A bison's consumption of the plant's carbon atoms in the form of sugar, starch, fat, cellulose, and resin, and then the animal's release of carbon dioxide. A truck's guzzling of fuel and oxygen and combustive spewing of ruptured carbon chains. Once I recognized fire as a process and not a thing, then sun, tree, animal, and machine became equally ephemeral. All were transmuting.

What was the first spark? The first ignition?

"It's got to be a lightning strike," said Dianne Edwards. Edwards was a luminary in her field, a distinguished research professor of paleobotany, the second female president of the Linnean Society, and a Fellow of the Royal Society. She spoke to me over the phone from her home in Cardiff, Wales. For six decades, she had been driven by her curiosity about the origins of plants and life. Now I asked Edwards how she had discovered evidence of the oldest known wildfire on earth, which burned 420 million years ago.

"It was just a pile of stones on the side of the road," she explained. In the early 2000s, Edwards was in the market town of Ludlow. She had grown up not far from there in the Welsh borderlands. Since the 1830s, scientists had mined what is called the Ludlow Bone Bed for evidence of the deep past—the spines, teeth, and scales of ancient fish and strange extirpated creatures. Edwards wasn't really interested in those, however. "The plants I focus on are not in the bone bed; they're in the finer-grain silt stones," she explained. So Edwards drove to the corner of a main road just south of town and took a couple of kilos of stones. "A fistful can keep you going for years," she said.

Back at her laboratory, Edwards macerated the rocks and dissolved them in hydrochloric acid, and when she looked at the remains under a microscope, she found beautifully preserved fossil fragments of an ancient plant she decided to call *Hollandophyton colliculum*. "*Hollandophyton* has never been found anywhere else. We were lucky to find it," she said. "It's just a thrill when you break open a bit of rock and you find something that's been there for four hundred million years."

Hollandophyton, with its simple forking stems bearing capsules at their ends, turned out to be one of the first plants on earth. The closest thing to it today would probably be found in the carpets of moss that cover a place like Iceland. The ancient plants were very tiny. "I reckon if you walked through a landscape with *Hollandophyton*, they would barely brush your toes," said Edwards.

The specimen had other fascinating qualities. Through the use of both light microscopy and scanning electron microscopy, Edwards saw that it possessed a black, silky luster, brittleness, and high reflectance. She realized that she was looking at the remains of organic tissues that had undergone chemical and structural changes as a result of charring. Charcoal, the product of fire, is resistant to decay and compression and can therefore preserve plant cells in three dimensions. In some places, paleobotanists have found preserved in exquisite detail flowers, fruits, seeds, and other plant organs that once burned. Now Edwards had evidence of a wildfire, most likely a low-temperature, smoldering one, that had burned some of the earliest plants on the planet. "These plants are very small, so it's difficult to imagine how there would be much to burn," said Edwards. "I think the plants accumulated in channels [of water], and then, when you had a period of drought, everything dried out, and you've got a source for combustion." Edwards emphasized the serendipity of such an event: plants dry enough to combust, lightning that struck in just the right place and moment. And, most importantly, enough oxygen in the atmosphere to spark the chemical reaction necessary for plant material to be transformed into heat, light, and gases. "You can't have fire if you don't have a certain level of oxygen in the air," she said.

THREE BILLION YEARS ago, the earth's atmosphere was mostly carbon dioxide. Then the cyanobacteria appeared, little blue-green pearls turning light and water into energy and burping out so much oxygen that they launched the Great Oxidation Event, which spurred other vegetal creation. The greening of the world. *Viriditas*. The twelfth-century abbess Hildegard von Bingen saw the "greening green" as the vegetal power of creation, God in plant flesh, and plants as slow fire that contained the heat and light of the combusting sun. She wrote:

> *O most honored Greening Force,*
> *You who roots in the Sun;*
> *You who lights up, in shining serenity, within a wheel*
> *that earthly excellence fails to comprehend.*
> *You are enfolded*
> *in the weaving of divine mysteries.*
> *You redden like the dawn*
> *and you burn: flame of the Sun.*

The earliest tiny, leafless, vascular plants provided two of the critical ingredients for fire: oxygen and fuel. And lightning, as Edwards had pointed out, was the likely first ignition. But volcanoes and asteroids could also ignite fires. Some scientists think that the asteroid that erased three-quarters of all life on earth, including the dinosaurs, may have ignited a wildfire encompassing the globe. The ejecta—the mass of particles blasted outward from impact—might have produced so much kinetic energy that it heated the upper atmosphere into incandescence, compressing and warming the air until it became light, causing infrared radiation to set the world ablaze.

There are more recent instances of asteroids becoming ignition sources. On the morning of June 30, 1908, a rock about 120 feet across entered the earth's atmosphere over Siberia at a speed of 33,500 miles per hour. According to NASA, the asteroid weighed 220 million pounds and heated the air surrounding it to 44,500 degrees Fahrenheit. At 7:17 a.m. the rock exploded into a fireball. The

First, Lightning

energy released was equivalent to 185 Hiroshima bombs. It took nineteen years before a team of scientists could reach the blast area of what was by then called the Tunguska asteroid. What they found was eighty million trees blown over in a radial pattern, and evidence of wildfires more than a hundred square kilometers from the center of impact. The fires had generated a smoke plume so high that the aerosols it pumped into the atmosphere brightened the night skies over Europe. "The sky was split in two, and high above the forest the whole northern part of the sky appeared covered in fire," said one witness.

A couple of years before I went to Nebraska, I became the steward of forty acres of woods in a little hollow between two mountains in Catskill State Park, in southeastern New York. An old logging cabin stood on the land, whose diminutive mountains and craggy escarpments had been created by the uplift and erosion of the Acadian orogeny, a geologic event that began hundreds of millions of years ago. The forest and creeks overflowed with Devonian-era stones tumbling down the mountainsides. I often picked up these rocks and imagined the specks of ancient wildfires they might reveal.

One of the earth's oldest forests was less than an hour's drive from my cabin, near the town of Gilboa. I couldn't see this primordial forest because it lay under a pile of gravel that the state highway department had put there to protect the fossilized root systems of the trees and hide them from potential looters. The evidence of the forest had survived for half a billion years, but today it is what the paleobotanist William Stein, a professor at Binghamton University, described to me as an "ephemeral site."

No one would even know that the ancient forest existed if it were not for the disturbances—hurricanes, floods, quarries, dams, and tunnels—that periodically exposed the sediment's secrets. "The best thing to happen to paleobotany was we blew up the Catskills,"

Stein said. He compared the discovery of the fossilized forest near Gilboa to the botanical equivalent of finding dinosaur footprints. It was first exposed in the late nineteenth century when workers quarried rock to build the Gilboa Dam. As they dug, they uncovered dozens of fossilized tree stumps in the inky blue, anoxic stone. The stumps were some four feet in diameter with radiating roots that ran nine feet long. In the 1920s, the first woman paleontologist for the state of New York was Winifred Goldring. Often photographed in pantaloons or holding her rock pick in the sidecar of a motorcycle, Goldring spent years studying the specimens. She determined that they had constituted the oldest forest on earth, which was made up of *Eospermatopteris*, a genus of giant, fern-like trees that grew along the swampy shore of a primal sea. Goldring imagined the experience of being in such a forest as strangely hushed, writing in one scientific paper:

> There could have been no dense shade in this primitive forest; except perhaps for the heavy moist atmosphere sunlight could easily filter through. No higher forms of life existed there. The hum of insects was not heard, for there were no insects here at this time. All the sounds one would hear could one have been in that ancient forest would be the murmuring of the winds in the tree tops or sounds from the neighboring sea or at times the howling of destructive storms.

Hundreds of millions of years later, another storm struck the Catskills. Hurricane Irene destroyed roads, and the backfill in Goldring's old quarry was needed to repair them. The state contacted Stein and his colleagues and offered them two weeks to reexamine the quarry floor before it would be covered up again. In addition to *Eospermatopteris*, Stein identified two more plant species, one of which was a vine-like plant with small photosynthetic stems that was the first

First, Lightning 47

on earth to produce a cambium and wood. This forest, they realized, was only the second-oldest forest on earth. In another quarry thirty miles away, Stein and his colleagues found spectacular trace fossils of root systems belonging to trees that lived 386 million years ago, a couple million years older than those at the Gilboa site. There, in the dusky red siltstone, mottled bluish grey in places, three different types of trees with roots spreading radially, some of them thirty-six feet long, had once grown. Among them was a tree called *Archaeopteris*, the first known plant to form leaves, and a so-called scale tree belonging to *Lycopsida*, the oldest group of living vascular plants.

The ancient forest likely had well-drained soils and wet and dry seasons. In those periods of dryness, the crackle of fire probably joined the sounds of wind in the trees: core samples collected from the quarry floor contain specks of charcoal. The emergence of forests over several million years during the Devonian period changed the surface of the planet. They made the earth flammable. "Everything hinges on their origins," said Stein.

HOMO ERECTUS LIKELY started using fire as far back as 1.8 million years ago. Some archeologists believe that this represents the most important technological innovation in human evolution. In South Africa there is a cave known as Wonderwerk that contains the remains of a million-year-old cooking fire a hundred feet from the entrance. Four hundred thousand years ago, human pyrotechnology seems to have swept the globe from sub-Saharan Africa to the Middle East to Europe. An amazingly diverse number of hominin subpopulations populated the earth at that time. *Homo sapiens* was there, as well as *Homo naledi*, Neanderthals, Denisovans, and *Homo erectus*. All of them used fire.

In places like Israel's Qesem Cave, ash-rich deposits from permanent hearths that are 400,000 years old run fifteen feet deep. At

Bolomor Cave in Spain, there are fourteen hearths that were consistently used from around 350,000 to 100,000 years ago. Permanent hearths likely created what archeologists have described as a change in the spatial organization of human communities. The hearth became an "evolutionary arena for social interaction." Others view the evidence of geographically widespread patterns of regular fire use as one of the earliest examples of cultural diffusion—the spread of a technology between humans over a large region. In a recent paper for *Proceedings of the National Academy of Sciences*, a group of Dutch archeologists argue that widespread fire use challenged the idea that hominin populations were isolated from one another by geographical barriers. Instead, both fire and new genetic insights indicate that these populations encountered and interacted with one another repeatedly. They likely tolerated one another, were perhaps even friendly, and shared their fire skills.

Someone must have been responsible for feeding fires while others left to hunt and forage. Permanent hearths might even have been the origin of our concept of home. The anthropologist Michael Chazan has speculated that these were places that one left and came back to and where food was shared. They may even have had spiritual significance: some permanent hearth sites also became burial sites. In Iraq's Shanidar Cave, the remains of ten Neanderthals were buried with flowers and chert stone points close to a large fire that once burned there. Fire, as Chazan has pointed out, is a chemical reaction that escapes our normal experience of the "tactile and real," veering on the magical. But that doesn't mean our ancient ancestors didn't understand it.

> Hunting would require an understanding of the behavior of animals; collecting would have required similar knowledge of the processes through which plants grow and multiply. The simple

act of making a stone tool involves the propagation of fracture, an act that irreversibly ruptures chemical bonds at a speed that cannot be measured. Fire is extraordinary, but early humans were familiar with the extraordinary and were adept at manipulating forces that cannot be seen and that modern science still struggles to understand.

By the Upper Paleolithic period, starting roughly 50,000 years ago, our ancestors were carrying portable lamps across Europe and likely lighting fires on the landscape to help them hunt, travel, and forage. While it can be difficult to distinguish between wildfires and human ignitions, a lot of evidence indicates that wherever humans went they ignited their surroundings. The charcoal record in New Zealand, for instance, shows that wildfires were likely occurring about twice per millennium in the island's temperate forests. Then, 800 years ago, people arrived and began setting fires so frequently that there was a loss of forest cover. In Africa, people may have been altering natural fire regimes as early as 40,000 years ago, and the area of land that burned annually on the continent probably peaked 4,000 years ago. In Alaska, burning increased sharply around 14,000 years ago, when humans are believed to have arrived there. There is evidence of anthropogenic burning even in places that were thought to be impervious to fire, like tropical rainforests. At the end of the Pleistocene epoch, also known as the Ice Age, charcoal from human fires spread from east to west across most of South America. Humans were transforming the world.

MY WOODS WERE dominated by beech, eastern hemlock, and maple trees. It was a "climax" forest, meaning the natural succession of plants over time had reached its final stage and a state of relative stability. A long time had passed since it had been disturbed by fire, but there was evidence of the forest's past everywhere, if you just knew where to look. A few miles away, for instance, sat the stumps of old American chestnut trees, *Castanea dentata*, the giant, mast-bearing trees that were the "sequoias" of the east. Black and white photos from the turn of the nineteenth century show settlers standing next to chestnut trees ten feet in diameter and over a hundred feet tall, the ruts in the bark so deep you could fit your whole hand between them. Before they were driven to functional extinction by blight in the twentieth century, American chestnuts grew across two hundred million acres of land stretching from southern Maine to northern Louisiana. Fire was common through the American chestnut's range. Samples of the trees' cores show that they burned at least every twenty years but sometimes as frequently as every two years. The chestnut was a pyrophyte, meaning it had traits that were conducive to fire. Its leaves were quick-drying and highly flammable and created a bed of tinder on the forest floor that would have burned and carried flame easily. The tree's branches, meanwhile, grew high up the trunk so that no branch could bring fire up into the tree's crown. In the wake of a fire, the chestnuts would have sprouted prolifically. This fire regime—the frequency of wildfire as a result of climate, vegetation, and ignitions—created an open-stand structure of large, mature trees, with the more shade-tolerant juveniles growing up in the wake of the flames.

Chestnut trees grew in the Catskills, however, because people wanted them to—the old stumps are a human artifact. According to naturalist Michael Kudish, in the higher regions of the mountains, northern hardwoods like beech, maple, yellow birch, and hemlock naturally dominate. It was the Haudenosaunee, or Ongweh'onweh, the Iroquoian-speaking confederacy of First Nations peoples in the Northeast, who repeatedly burned the land to extend the range of chestnuts and other nut trees like walnuts and hickories. Fire-loving species like red, black, scrub, and white oaks, shagbark, pignut, and butternut hickories, sassafras, black birch, blueberry, huckleberry, mountain laurel, mapleleaf viburnum, whorled loosestrife, and pitch pine flourished in burned areas. Kudish thinks it is possible that some 95 percent of fires in the Catskills were started by people; the environment was just too wet for lightning fires to carry very far. Thanks to these ignitions, the forest was a nut orchard, feeding people and animals while regenerating itself. Even early European colonists, according to Kudish, saw the benefit of burning the forest.

When chestnut trees disappeared, the shade-tolerant, mesic tree species moved in, inhibiting ignition with their higher fuel moisture. Oaks are also pyrophytic and once benefited from the combustible forests created by American chestnuts. Oaks love sun, and they flourished in environments that underwent periodic, low- to moderate-intensity surface fires that kept the forest canopy open and reduced the competition from other trees. Before colonization, oak woodlands once covered as much as 70 percent of what is now the central and eastern United States, and their mast—the edible acorns—were the foundation of a vast food chain. Without fire, oaks have fewer chances to regenerate. From 1980 to 2008, 60 percent of eastern forests underwent a decrease in oak density. This has created what biologists call an oak bottleneck: once the remaining overstory of oaks dies, there will be few oaks to replace them. The whole forest will become even less flammable.

At my cabin, I often wrote in an old outhouse constructed of a shipping crate that had once transported periodicals from the Vatican Library to New York City. I would stare out the door into what is now an asbestos forest: a place so moist and dominated by northern hardwoods that it is considered impervious to wildfire. But I could see a tall, single, majestic oak from my vantage, its crown as elegantly splayed as a ballerina's arms. The tree had to be at least 150 years old—maybe it had once felt a fire dancing across its roots?

THE NATIVE AMERICAN theologian and scholar Vine Deloria Jr. wrote in *Red Earth, White Lies*, "People want to believe that the Western Hemisphere, and more particularly North America, was a vacant, unexploited, fertile land waiting to be put under cultivation according to God's holy dictates.... The hemisphere thus belonged to whoever was able to rescue it from its wilderness state." Most people still accept that wilderness is a real thing. In their image of North America before colonization, it was a wild place where hunter-foragers lived in a benign and idyllic balance with nature. This belief is why, when people visit a national park, they see it not just as preserved geographical space but also as preserved time. They go to see the past.

I was dispelled of these ideas by the work of the anthropologist Omer Stewart. Born in 1908, Stewart was a Mormon from Utah whose ancestors converted to the religion at its very beginning. He was an elder in the church and went to Switzerland as a young missionary. During the Great Depression, Stewart decided to study law and happened to enroll in an anthropology course that led to a summer job as a camp cook on an archeological dig. Stewart ended up helping to excavate a prehistoric Shoshone site in Utah called Promontory Cave, and the experience changed him. He decided to study anthropology full time. After two years he had completely lost faith in what he considered the dogma of Mormonism. He was excommunicated from the Mormon Church and moved to Berkeley, California, where he worked as a dishwasher so he could study under A. L. Kroeber, one of the founders of cultural anthropology.

Despite Stewart's atheism, or maybe because of it, he became a fervent advocate of religious freedom. He dedicated most of his anthropological career to the study of peyote and its use by the Native

First, Lightning

American Church. At the time, the church had around two hundred thousand members, and peyote was the church's divine sacrament, taken to communicate directly with the Creator. Stewart became a participant-observer in the church's practices, using peyote and drumming until his fingers blistered. "The pleasure and enlightenment I felt, in spite of an all-night ordeal of gastric upset and vomiting," he later wrote, "was impossible to explain." He also became an expert witness in court cases attempting to protect the practices of the Native American Church as an expression of religious freedom.

Stewart's other scholarly interest was fire. One of his first professors at Berkeley was the geographer Carl Sauer, who helped develop the idea of cultural landscapes, places that were shaped by cultural forces. Sauer was also interested in fire: he had studied the Ozarks of Missouri as a graduate student and noted how Native Americans had set fire to the grass yearly, a practice that was then continued by many settlers. In 1935, Kroeber enlisted Stewart to create an ethnogeography of the Pomo Indian tribes in California for a salary of $100. Stewart was to "take a pencil and notebook" and travel north of Berkeley to interview "the oldest Indians [he] could find." Stewart ended up interviewing dozens of people and creating maps that delineated the different tribes and subdivisions of the Pomo. His research became important documentation for subsequent land-claims cases in the state. During one such conversation, a man pointed to a hill that was covered in mesquite and told Stewart that it had once been a place covered in grass where animals would feed. Now, Indians weren't allowed to burn anymore, and the man thought the country was ruined. Stewart realized he had stumbled on the "ancient practice of intentional burning, setting fires to the landscape every year to clean out the brush and open up the forest so game could be spotted."

After this conversation, Stewart began, as he described it, "gathering nearly every reference in existence" to intentional burning by

Native Americans. The evidence he amassed that Native Americans not only imitated natural fire regimes but manipulated them to serve their cultural interests was overwhelming. His research eventually grew to hundreds and hundreds of pages of sources, first-person accounts, and data. He cited over two hundred examples of cultural fire traditions. Native American societies, Stewart realized, had shaped the geography and landscape of North America. Not only was intentional burning a universal pattern across the continent; he reasoned it had probably been practiced the world over.

Many of the historical accounts Stewart gathered are fascinating. As early as 1609, English settlers were reporting fires burning for weeks at a time. In Virginia the smoke was at times so pervasive that the mountains became invisible. They described forests free of underbrush so that a deer could be spotted through the timber as far away as the eye could see. One settler wrote of the forest floor as a field of berries, so many that "your foot can hardly direct itself where it will not be dyed in the blood of large and delicious strawberries." In these woods, a horse could be ridden at full speed without risk of running into a tree.

In 1632, the colonist Thomas Morton wrote:

> The Savages are accustomed, to set fire of the Country in all places where they come; and to burn it, twice a yeare, viz at the Spring, and the fall of the leafe.... For when the fire is once kindles, it dilates and spreads it self as well against, as with the wind; during continually night and day untill a shower of raine falls to quench it. And this custome of firing the Country is the means to make it passable, and by that meanes the trees growe here and there as in our parks: and makes the Country very beautiful and commodious.

In the early 1900s, the forester Hu Maxwell wrote, "The Indian is by nature an incendiary, and forest burning was the Virginia

Indian's besetting sin." A settler traveling up the Potomac River in 1633 described how "on each bank of solid earth rise beautiful groves of trees, not choked up with an undergrowth of brambles and bushes, but as if laid out by hand, and in a manner so open that you might drive a four-horse chariot in the midst of the trees."

In North America's seas of grass, settlers found a similar situation: fires were common and often intentionally set. In 1832, the artist George Catlin tried to capture the image of a prairie fire: giant coils of black oil paint represent smoke filling the sky, and dozens of Native Americans outrun the flames on horseback. He called the fires "thunder rumbling as it goes." Ten years later, Catlin still couldn't seem to erase from his mind the image of the fires he witnessed:

> The prairies burning form some of the most beautiful scenes that are to be witnessed in this country, and also some of the most sublime.... Over the elevated lands and prairie bluffs, where the grass is thin and short, the fire slowly creeps with a feeble flame, which one can easily step over... where the wild animals often rest in their lairs until the flames almost burn their noses, when they will reluctantly rise, and leap over it, and trot amongst the cinders, where the fire has past and left the ground as black as jet. These scenes at night become indescribably beautiful, when their flames are seen at many miles distance, creeping over the sides and tops of the bluffs, appearing to be sparkling and brilliant chains of liquid fire (the hills being lost to the view), hanging upended in graceful festoons from the skies.

Stewart believed that America's grasslands would not have existed on such a scale had Native Americans not repeatedly burned them. Over thousands of years and millions of acres, he argued,

Native Americans had even influenced the prairie climate itself: its ground temperature, evaporation, wind, and precipitation. If it weren't for their fires, juniper, ponderosa pine, hackberry, sagebrush, and mesquite would have taken over. "The extensive grasslands that we call prairies and plains owe their existence to fire set by Indians and continued by accident or design to the present," wrote Stewart.

The manuscript that resulted from Stewart's scholarship was eight hundred pages long. He called it *The Effects of Burning of the Grasslands and Forest by Aborigines the World Over*. But it was never read by the public. First, his research was interrupted when he was conscripted to fight in World War II. Then, in 1952, the publisher Alfred A. Knopf dismissed it as having insufficient appeal to readers. Stewart next submitted the manuscript to a research institution, where it went missing for twenty-five years.

Stewart's scholarship was radical. The idea that Native Americans had had extraordinary impacts on the world around them contradicted fundamental ideas that they represented an earlier and more primitive stage of human evolution and lacked the ability to impact their environment. Even Kroeber, Stewart's mentor and teacher, believed that culture was informed by geography and that geography could not be determined by culture; that Indigenous people were essentially passive, foraging and hunting foods from pristine environments. Stewart was a heretic offering a revisionist history.

The ethnoecologist M. Kat Anderson has pointed out that if ecologists and environmentalists accept the premise that people shaped the ecology of plant communities with fire, they are forced to reconsider wilderness philosophy itself. And if they set down this path, Anderson writes in her book *Tending the Wild*, they have to

> face up to the removal of Native Americans from wilderness areas as in at least some instances a grave *faux pas* that would

ultimately undermine the unique habitat types and the biological diversity that they sought to preserve. They would also have to reevaluate the assumption that land use and conservation are always incompatible or that human tinkering with nature is inevitably destructive.

The historian Stephen J. Pyne told me that stripping people from landscapes made for easier scientific modeling; accepting human influence on the landscape would have made models so complex as to be unworkable. For these reasons, scientists begin with untainted nature, then add people as a disturbance. They start with simplicity and move toward complexity. "An anthropogenic model turns that thinking on its head," he said. "Science and wilderness formed a strong nuclear bond."

Stewart particularly regretted that his own peers—anthropologists—ignored evidence that Indigenous people were powerful agents who had transformed the physical world. "An abundance of evidence of the degree to which the aboriginal population had influenced and *altered* the environment was being systematically overlooked to the detriment of all science," he lamented. In the early 1990s, Stewart submitted his book manuscript one last time. It was accepted by the University of Oklahoma Press. He passed away less than a year later and never saw the book that was eventually published under the title *Forgotten Fires*.

OMER STEWART TURNED to the historical record to find evidence of anthropogenic burning. In later decades, scientists have turned to another source: trees themselves. When fire burns a tree, the heat kills the cambial cells—the layer of the trunk that grows each year. So long as the tree isn't killed, these dead cells are eventually covered with new growth that hides the scar. By sampling a tree, scientists can reveal these scars and determine not only what year the fire burned but even how hot it was and what direction it came from.

Fire scars provide dendrochronologists—scientists who study growth rings of trees—with a record that can span millennia. The oldest recorded fire scar was discovered in a giant sequoia over three thousand years old. I once stood before a slab of a giant sequoia hanging on the wall of the Laboratory of Tree-Ring Research in Arizona. It was polished to reveal in greater contrast the pattern of its growth rings, and arrows were painted in white indicating the years it had burned, with the date written next to each. The tree had experienced its first fire in AD 561 and its last in 1915. I counted nearly a hundred and fifty in between. Scientists have found fire scars in all corners of the continent across nearly every ecoregion, from the Blue Mountains of eastern Oregon to the Ozarks of Missouri, and in over ninety different tree species. The North American tree-ring fire-scar network has compiled a database of thirty-seven thousand fire-scarred trees from 2,562 sites, each one of which contains the year of the first and last fire scar. By looking into patterns and variability in this data, scientists can explore the climate and geographic drivers of fire regimes over long time scales. It also provides a record of the way that humans have impacted fire regimes throughout history.

In the Jemez Mountains of northern New Mexico, the dendrochronologist and forest ecologist Tom Swetnam has been using fire-scarred tree rings in addition to charcoal and sediment samples to reconstruct the fire history of the Jemez people, who have lived there for at least a thousand years. The former director of the University of Arizona's Tree-Ring Lab, Tom described the results of his research as evidence that the Jemez Mountains have been a "humanized landscape" for centuries. To protect villages from fires, the Jemez people likely created defensible spaces by removing pine needles and combustible fuels within a half mile of their homes and cutting lots and lots of small trees. "The Jemez people were undoubtedly interacting with fire in really, really important ways," he told me. They set many fires, but the fires were small in scale. "When people were here, there were actually *more* fire events, but they were not extensive. They were patchy," said Swetnam. "We see this in the tree-ring record. There are just lots of fire scars before 1650, but they are asynchronous between the trees. There's a fire here one year and the next year there's a fire a hundred feet away. Lots of events, but they're little."

When the Spanish arrived in the early 1600s, the Jemez people were decimated by disease and subjugation; within a few decades, the population had crashed by 80 to 90 percent, according to Swetnam. People were forcibly moved out of their villages and field houses into the valley to live near missions. Without people, the forest began to change, and the fire scars tell this story, too. "After 1650, the overall fire frequency is not that different, but the fires are extensive," said Tom. "Every three to ten years—boom! Everything burns. The total *area* burned hasn't changed, but the spatial configuration of fire completely changes without people."

Even temperate rainforests contain fire-scarred trees. On Hecate Island in British Columbia, a place that swirls in fog and 160 annual inches of rain, scientists have been able to reconstruct seven hundred years of fire activity. Alongside sites of clam gardens and fish traps, scientists took tree sample cores and found ninety-nine fire scars dating from 1376 to 1893; lightning is so rare here that it cannot explain the frequency. The findings indicated that people burned near habitation sites in order to encourage growth of the western red cedar, a culturally significant tree used for clothing, medicine, canoes, and homes. The fires also likely stimulated growth of important plants and served to maintain open canopies where berries could proliferate.

Scientists call landscapes that were historically maintained by Indigenous people anthropogenic ecosystems. Forest orchards, garden agriculture, mariculture, meadows, and prairies are all examples of such places. To make them productive, people pruned, broadcast seeds, transplanted, fertilized, dammed, weeded, tilled, and irrigated. And they burned. On Whidbey Island in Washington there is an anthropogenic meadow called Ebey's Prairie. By analyzing soils and artifacts, scientists have determined that people had been using the prairie for at least ten thousand years and setting fire to it for at least twenty-three hundred years. In Oregon's Willamette Valley, Garry oak ecosystems are the result of frequent human-ignited burning for the last six thousand years.

In Canada's boreal forest, a forester once claimed it was possible to follow the "course of an Indian travelling through the woods" from a distance "by the succession of smokes as he set fires." Anthropogenic burning in the region was purposefully situated to create what anthropologists describe as "yards" and "corridors." Fire yards were openings like prairies, meadows, and swales surrounded by forest; fire corridors were burned areas along streams, trails, and ridges. By cyclically burning these areas, people could manage complex interactions and relationships between plants and animals. After a burn, mice and rabbits would proliferate in the new grass, attracting fox and lynx; white poplar would sprout for the beaver; moose would come to eat new leaves; muskrats and ducks would come to nest and feast on fresh roots. When it was too wet to burn, people often made a big campfire and then left it to smolder for weeks until the grass was dry enough and the fire could do its job.

At the time of European settlement, what is now California had the greatest number of diverse Native cultural groups for an area its size in the entire Western Hemisphere, and anthropogenic ecosystems were everywhere. Ecologists estimate that three-fourths of the state's vegetation is influenced by fire, and that as much as thirteen million acres once burned annually from human and lightning ignitions. "Like other great spiritual and philosophical traditions

First, Lightning

around the world," wrote M. Kat Anderson, "California Indians recognized that life is dynamic, relying on cycles of destruction and renewal." In places like the Klamath Mountains, the burning practices were varied and complex and supported multiple cultural purposes. As Bill Tripp, director of the Karuk Tribe Department of Natural Resources, explained, fire was utilized to create a traversable system of trails along the mountain ridges. Both lightning strikes and tribes kept the ridges so open that the surrounding forest looked like a "pruned orchard." In the fall, people brought fire down from the ridge systems and burned around villages and camps and ceremonial sites, maintaining the oak woodlands, prairies, and mixed conifer forest. Traditionally, Tripp said, women burned to fulfill their responsibility to plants, and men burned to fulfill their responsibility to animals.

Elizabeth Azuzz, a Yurok tribal member and the secretary of the California nonprofit Cultural Fire Management Council, describes her tribe's work to practice their fire-lighting knowledge and traditions in northern California as "tending the land." When she was four years old, her grandfather found her lighting the backyard on fire and asked her to put it out. He then explained that, as a Yurok, she had a duty to light the fires that would generate basketry materials, destroy the weevils that ate acorn crops, keep the creeks that salmon ran through cool with smoke, and thin the forest floor for grazing elk. The goal of fire-lighting was to maintain the natural resources critical to her culture's survival. "Fire is life for us. Fire is family. It's a tool that we use to be able to restore our environment, our ecosystem, and maintain the strength and health of our people," she said.

Azuzz was the first person I ever heard explicitly describe a deep affection for fire. "I love fire. I love the way it smells. I love the way it looks. I love the job," she told me. "I have been called a pyro my whole life. I actually relish that term because, yes, I am. It's just a part of who I am. I was always raised to carry water and fire." Azuzz has raised her own grandchildren around fire and passed down her grandfather's message of their duty to the land. "We instill it in our children because we could be deemed arsonists if we are caught in a situation that others deem unnecessary," she explained. For over a century, the Yurok have been imprisoned for setting fire to their land, even though they consider it their God-given right. Members of Azuzz's own family have been incarcerated for lighting fires.

As for herself, Azuzz refuses to get the governmental qualifications that wildland firefighters are typically required to maintain in order to conduct prescribed burns. "I do not believe that suppression firefighters have the right to tell me that I don't know what I'm doing," she said. "They are not cultural burners. All they do is put things out. All I do is light things."

3 PYROPHILIA

I THOUGHT I MIGHT LIGHT A FIRE IN NEBRASKA AND MOVE ON, BUT I WAS WRONG. Something had changed. Fire reoriented my perspective. I came back from Ord and walked in a forest or field imagining how it would burn in different winds. As the spring melt at my cabin revealed leaf litter and fallen branches, I thought, Wouldn't it be nice if I could clean this all up with some fire? I wandered the asbestos forest thinking of nothing but flame.

To other wildland firefighters, there was a clear explanation for what had happened to me. "You have the fire bug," said Jeremy Bailey, the director of prescribed-fire training for the Nature Conservancy, a nonprofit environmental organization. Bailey lived in Colorado and helped me navigate the often obtuse wildland firefighter-training qualification system. Over time, he became my guide to the greater fire community—its history, culture, and issues—as I sought to become a kind of participant-observer on fire crews.

Bailey sported a black beard and coal-colored eyes and had a serious, even stern, demeanor. But no one was more generous with their insights and advice. It was Bailey who had come up with the

TREX concept: an event where wildland firefighters and practitioners gathered to burn and learn together. Bailey admitted that he had predicted what would happen when I went to Nebraska; he had seen it many times before. "You were going to fall in love with it, you were gonna want to keep going back, you were gonna see all this stuff," he said. "We get bitten by the fire."

The term "fire bug" was common among wildland firefighters. Some had been bitten by it, while others caught it like a virus. Some described the fire bug as something that had always existed in them. Bailey's own passion for fire had begun in 1995 when he worked as a wildland firefighter and continued after he joined the Santa Fe Interagency Hotshot Crew. For part of the season, his crew worked in the Jemez Mountains in New Mexico, igniting controlled burns to improve the health of ponderosa pine forests. "Almost immediately, I had this awareness of the need for fire," Bailey told me. But as time passed, he noticed a pattern of federal agencies prioritizing suppression over ignition. Every year, he would read what's known as a Moses Letter—a sort of wildland-firefighting State of the Union statement that the US Forest Service chief sends to the supervisors of federal firefighting employees. The letters always seemed to say the same thing: let my people go fight fires, because this year things are so bad that we need to suppress every single one. "I clearly saw the writing on the wall," Bailey said. In 2008, he left firefighting to work for the Nature Conservancy to begin building a national network of good-fire practitioners.

Through Bailey, I got word that a temporary crew was assembling in Montana to burn five hundred acres in the Blackfoot Watershed east of Missoula. The crew would sleep in tents for a couple of weeks, and food would be provided. I signed up and bought my airline ticket. It was a forest of fire-dependent ponderosa pine and western larch; I had never burned in timber before, and I was nervous, excited, eager to work. But it was an unseasonably wet and cold spring in Montana, and the burn got canceled.

I started trying to find a place to burn in New Jersey, which was closer to home. Every year, the state burns around twenty-five thousand acres of inland pine barrens, places with sandy, acidic soil, orchids, carnivorous plants, and a dependence on fire. Pine barrens are rare today because they were considered wastelands for so long. People bulldozed them for development or suppressed fires and allowed them to turn into deciduous woodlands. Despite this, the state still contains the largest inland pine barren ecosystem in the country. In some people's opinion, it is also home to one of the longest uninterrupted fire-lighting cultures, beginning with the Lenape. At the time of colonization, the "sweet perfume" of forest fires could be smelled at sea long before the land itself was visible. In 1632, the Dutch navigator David DeVries wrote, "This [smell] comes from the Indians setting fire at this time of the year, to the woods and thickets, in order to hunt; and the land is full of sweet-smelling herbs, as sassafras, which has a sweet smell." But the spring I was looking for work, the state shut down burn operations early. A couple of aggressive wildfires fed by forty-mile-per-hour winds and low humidity had made it too risky to burn longer. I was like the pyro Goldilocks: Montana was too wet and cold, New Jersey too dry and windy.

My efforts to find a crew became slightly deranged. I would wake up and open InciWeb, a national database listing current wildfires and controlled burns, scanning the map and clicking on a prescribed burn to see details of the fire, its acreage, and the managing agency, and then I'd reach out to the overhead, asking if I could come volunteer. I got a good tip from someone who had spent the weekend burning in Michigan's Sleeping Bear Dunes National Lakeshore. "One of the coolest burns I have ever done," he wrote.

"Burning oak pine savannah in dune swale. They have more plans to burn later in the week." I reached out to the burn boss, but when we spoke on the phone, he sounded crestfallen. He was sitting in the cab of his pickup truck, watching the rain fall over the dunes. "It's raining so much here," he said. "I don't know if we'll even get one more day in." We chatted to pass the time—or maybe I hoped if I kept him on the phone long enough, the sun would come out and he would tell me to get on a plane.

Finally, I got lucky. I was referred to the fire manager of something called the Albany Pine Bush, a rare pitch-pine scrub-oak ecotype in upstate New York. I wrote to him, and he replied. "I would be happy to talk to you about upcoming burns or the pine bush in general. I send out emails when we're looking to burn." When we spoke, he explained that the pine bush had once spanned forty thousand acres but was now reduced to three thousand, squished between freeways, houses, and malls. Prescribed fires at the bush burned just a few miles from both the state capitol and an international airport with the help of two seasonal workers and a group of volunteers made up mainly of state forest rangers. He offered to put me on the volunteer list.

One day in early April, I opened an email. "Good Morning," it said. "We are a GO for tomorrow's burn." The next day I woke in the dark and began a journey I would make many times, driving under the fluorescent lights of the Battery Tunnel, up the empty FDR Drive, across the George Washington Bridge, and then bombing up Interstate 87 as dawn broke. "Fire is a powerful drug," Zeke Lunder had warned me.

IN THE SPRING of 1887, the entomologist Samuel H. Scudder collected some tiny butterfly eggs in a place called Karner, New York. It was an area known to abound with Lepidoptera: spring elfins, dusky wing skippers, cobweb skippers, and inland barren buck moths. The train stop in Karner, which sat in the midst of a vast pine barrens, was called Butterfly Station. Scudder raised the eggs on a wild blue lupine plant, drawing a careful rendition in his three-volume tome *The Butterflies of the Eastern United States and Canada*. Each egg looked like a miniature sea urchin with a variegated surface and a central depression like a belly button. Scudder watched as pea-green larvae emerged and began to eat the leaves of the wild lupine he provided. After a few weeks, the larvae transformed into dark blue pupae, and each then emerged from its chrysalis as a small, shimmering jewel: blue sapphire males with undersides of aquamarine dusted in orange crescents. The females were duskier, the color of a stormy sea with iridescent azure streaks. He thought it was a subspecies and called the butterfly *Lycaena scuderii*, or Scudder's blue.

Before Scudder died from Parkinson's disease, he donated his collection of insects to the Museum of Comparative Zoology at Harvard. It was there, in 1943, that the novelist and lepidopterist Vladimir Nabokov found a male specimen that Scudder had collected and raised from an egg. But Nabokov believed the butterfly was actually new to science and gave it a different name: *Lycaeides melissa samuelis* or the Karner blue butterfly. The male specimen became the holotype of a new species, distinguished by a red rather than white label in the collection. Nabokov wrote of the experience in a poem:

> *Wide open on its pin (though fast asleep),*
> *and safe from creeping relatives and rust,*
> *in the secluded stronghold where we keep*
> *type specimens it will transcend its dust.*
>
> *Dark pictures, thrones, the stones that pilgrims kiss,*
> *poems that take a thousand years to die*

> *but ape the immortality of this*
> *red label on a little butterfly.*

Eight years later, Nabokov passed through the pine barrens near Karner where he thought the little blue butterfly "ought to be out." As he wrote in a letter to a friend, "Yesterday morning on our way back, we drove to a certain place between Albany and Schenectady where, on a pine-scrub waste, near absolutely marvelous patches of lupines in bloom, I took a few specimens of my little *samuelis*." The butterflies were "still doing as fine under those old, gnarled pines along the railroad as they did ninety years ago." In his novel *Pnin*, Nabokov described his protagonist seeing Karner's blues in the wild: "A score of small butterflies, all of one kind, were settled on a damp patch of sand, their wings erect and closed, showing their pale undersides with dark dots and tiny orange-rimmed peacock spots along the hindwing margins; one of Pnin's shed rubbers disturbed some of them and, revealing the celestial hue of their upper surface, they fluttered around like blue snowflakes before settling again."

The fact that there was any pine bush at all in the middle of an otherwise deciduous northeastern forest was itself a kind of miracle. It had been created and maintained by fire, which prevented it from turning into northern hardwoods or an oak and pine forest, in which fire is less frequent. As the natural historian Jeffrey Barnes explains in his guide to the Albany Pine Bush, fire removes shade and insulation from the soil surface and reduces its reflectivity, leading to higher daily temperatures and increased plant productivity and soil organism activity. The plants that survive in the barrens' nutrient-starved soil find incredible ways to do so. Some create symbiotic relationships with mycorrhizae—root fungi—in which the fungi get high-energy carbohydrates from the host plants and the plants get nitrogen, phosphorus, and water in return. There were other mutualistic relationships; for example, the larvae of the Karner blue butterflies secrete a honeydew collected and eaten by ants, who then protect the larvae from predators.

Native Americans had undoubtedly contributed to maintaining the pine barrens by igniting frequent fires. The academic Timothy Dwight believes the pitch-pine and oak forests in New England had been burned for more than a thousand years. In the 1640s there were fewer than eight thousand Europeans populating New York, then called New Netherlands. A Dutch colonist described how Native Americans practiced an annual custom of "bush-burning" in the pines of Rensselaerswyck, which includes the present-day Albany County. "Those fires appear grand at night from the passing vessels in the river, when the woods are burning on both sides of the same," he wrote. "Then we can see a great distance by the light of the blazing trees, the flames being driven by the wind, and fed by the tops of the trees. But the dead and dying trees remain burning in their standing positions, which appear sublime and beautiful when seen at a distance."

Eventually, settlers became less entranced by the barrens. In 1870, an Albany publisher wrote that the pine bush was as "forlorn, miserable and unsatisfactory a combination of sand, swamp and aridity, as the Union can produce.... It is a mistake in nature, an enormous fraud on the worms and bugs which usually get a living somewhere on ordinary bad land." Even so, settlement may have benefited the barrens for a time. A train line, the first steam passenger train in the world, was built between Albany and Schenectady and, as recounted by Barnes in the *Natural History of the Albany Pine Bush*, "showered the flammable brush of the pine barrens with sparks." Then, at the turn of the century, fires stopped. New York was one of the first states to regulate the use of fire for agriculture, and in 1911, its state foresters became part of a national effort, in conjunction with the US Forest Service, to protect public lands from fire.

When Nabokov visited the pine bush, nearly all of its fires were being put out and its grasslands and wetlands were shrinking. Young white pines were beginning to compete with pitch pines. Over the years, the ravines of the pine bush became a garbage dump where people threw their trash and old cars and appliances. Big swaths of the barrens were developed into housing: over fifteen hundred acres of pitch-pine scrub-oak ecosystem were run through with roads and yards that acted as fuel breaks, further suppressing fire. Another seven hundred acres were lost as hardwood forest and non-native plants invaded and competed with the shrubs, grasses, and trees for space, light, moisture, and nutrients.

By the 1970s and 1980s, many of the pine bush's diverse insects had disappeared. Noctuid moths and giant silk moths that feed on young pitch pine vanished. The regal fritillary that feeds on bird's-foot violets was gone. Some thirty types of insects were lost, and another fourteen rare species underwent severe population declines. The effect of fire suppression on the Karner blue butterfly was particularly catastrophic. It took just five years without fire for shrubs and plants to shade out the wild blue lupine its larvae needed. By 1992, the number of Karner blue butterflies counted on the preserve was just three hundred, and the species' population had declined by 99 percent.

Nabokov called the pine bush "a sandy and flowery little paradise," but by the time Dieter Zimmer went there in the late 1990s to write *A Guide to Nabokov's Butterflies and Moths*, the place was a "zone of truck yards, storehouses, shopping centers, roadside diners and sundry small business. Right there, between the old New York Central railroad track (still in use) and a six-lane thruway, is an enclosed area of about eight hundred hectares, the remains of what once was a 100-square-kilometer pine barren: sand dunes, scrub oak, isolated pitch pines, a few trails."

When the Karner blue butterfly was listed as a federally endangered species, ecologists estimated that a minimum of two thousand acres of pitch-pine scrub-oak ecotype needed to be maintained to provide the butterfly larvae with wild lupine. To bring back the "blue snowflakes," they needed to start lighting fires.

I TOOK THE exit off of Interstate 87 just south of downtown Albany and the Crossgates Mall, heading west past an elementary school and across New Karner Road until I reached a dirt trailhead and parked my car. I had graduated from the banana suit and had bought my own green Nomex pants and a "yellow." Now I stood outside my car holding the spotless shirt, so new it glowed, and remembered the warning I had gotten: never show up on a fire with a clean shirt. I looked around. There was no ashy ground nearby. Should I run into the trees and rub it in the dirt? The fear of getting caught faking was worse than the fear of being seen as a rookie—which I couldn't hide anyway.

I walked down the trail to where the burn boss was writing the operational assignments for the day on the side of a white trailer. As the crew briefing began, the burn boss explained we would be burning two continuous units, "Cupcake" and "Chainsaw," for a total of eighty-four acres that day. I looked down at the paper map. This was not prairie. I was familiar with burn units that were rectangles or squares with ninety-degree corners. These were amoebas with squiggly perimeters that intersected with public trails and butted against Interstate 90. "We've got a mostly sunny day with a max temp of sixty-six and a low RH [relative humidity] of 37 percent," said the burn boss. "Winds are northwest at three to seven. The anticipated fire behavior is that Cupcake has better fuels than

Chainsaw. There's a lot of *rubus*, but I don't think it's going to be super hot." Two holding teams would start at the southeast corner of the units and begin igniting along the perimeter in opposite directions, building black, while the ignitions team—seven people with drip torches—would walk through the interior of the units lighting strips of fire. "Any slopover that takes longer than half an hour to put out, we'll call in the fire department," he said. "I don't anticipate that happening today."

From behind me, a man chimed in. It was Neil Gifford, the conservation director of the Albany Pine Bush. "I just want to talk about the ecology of the burn," he said. "This is the best remaining example of inland pitch pine in the country with rare and unique birds. It used to be twenty-five thousand acres, and we're looking to reset the natural fire regime that supported these species. It was thanks to Indigenous burning that we have this viable pitch-pine scrub-oak barrens that can support rare species like lupine and the Karner blue butterfly."

"You didn't say anything about supporting ticks!" someone shouted.

"Fire kills ticks," said Gifford. "What more do you need to know?"

I was assigned to the ignitions team. Our firing boss trainee was a young and brash state forest ranger who instructed us to grab drip torches and follow him. We hiked for fifteen minutes on a sandy trail and then set the torches on the ground to unscrew the caps. We took the spouts out of the canisters and flipped them outward, then carefully screwed the caps back on, mindful not to cross threads and risk spilling fuel on ourselves. We were at the southeast end of the burn unit, an open, field-like setting with crunchy dried leaves underfoot and scattered young pitch pines. "We're making south-to-north passes across the unit," said the firing boss. "We'll take it slow at first, flanking fire off the line as they're building black, and then we'll rip it off when we get to the other line."

At 10:50 a.m., someone lit the test fire with a lighter. The leaves crackled and the fire fluttered a bit, and then it became eager and leapt upward. Suddenly the cirriform flames were four or five feet tall, and a white pall of smoke began to lift skyward. I understood the operational purpose of test fires—to observe fire and smoke behavior. But there was something about them that felt ritualistic. I noticed how all of our eyes were diverted to the same place at the moment of ignition, and it got a little hushed, like we were acknowledging the arrival of fire with our attention; or perhaps the fire, in its vibrancy, captured us. Either way, I felt a demarcation in time, a threshold that we crossed from no-fire to fire.

"I want to see big smiles today," shouted the firing boss as we began. "We're doing God's work!"

I lit my torch in the flames of the test fire and walked into the unit, each ignitor spread twenty feet apart and staggered from right to left in front of one another. The grass quickly became head-high patches of brambles. Within minutes, I was out of sight of anyone, bushwhacking alone with the torch in my left hand dripping flame in a trail behind me as my right hand pushed through the brush. Sometimes I saw the flash of a fellow ignitor's trail of fire or caught a glimpse of a helmet through the brambles. Otherwise, I had to trust that I was staying on a true course and that the ignitor to my left would not pass me, thereby sending fire back into my path. I couldn't see the firing boss; he was somewhere out front, choreographing us over the radio. Thorns clawed at my legs, arms, and face. As the timber thickened, I clambered over snags. I got caught in ropy vines, fire nipping at my heels. I pushed as hard as I could to break free, my heart starting to pound. It was hot. I dripped sweat. There was only one direction I could go: forward. So, I thought, this is what it's like to light in timber.

But it was breathtakingly, extraordinarily beautiful. The burning forest floor emanated a rich and almost velvety brown smoke, tinged violet and calamine pink. The smoke gathered and condensed as it was drawn upward, a twisting and shimmering, vaporous cloud bathing the trees. In places, shafts of sunlight streaked through the smoke, and the flames lapped at the pitch pines, turning their flaky bark into purple rosettes that glittered with charred reflectance. As we moved into heavier forest, the smoke turned the trees into spectral ghosts around me. The forest had become liminal, transformed.

We dragged torch for hours, finally tying up the unit on the northern edge of Cupcake in a ring of fire. I sat in the dirt of a trail and watched the smoke rise into the sky. On one of the slender pitch pines, the flames crawled twenty feet up, likely following a trail of secreted resin or a wound. What's it like to be a pitch pine reunited with your old friend fire? I wondered.

I was no longer as bothered by smoke as I had been in Nebraska. Now I took pleasure in it. I had begun to appreciate that the smoke, just like the flame, was serving a purpose. Studies of wildfires show that smoke doesn't necessarily block light; it scatters light, diffusing it by as much as a third, which means that a greater number of leaves receive it. In a haze, plants use available light nearly twice as efficiently as direct light. Smoke itself contains thousands of different chemical compounds, many of which scientists have only begun to identify. In 2004, scientists in Australia and South Africa discovered that when cellulose combusts, it produces compounds called butenolides, which can promote seed germination.

The word for smoke in the Aboriginal Noongar language is *karrick*; in 2008, the active butenolide substance in smoke was named karrikinolide by an Australian molecular scientist, Gavin Flematti. The gene encoding the karrikin receptor is so ancient it can be traced back to algae and bacteria. So far, six karrikinolides have been identified. Flematti hypothesized that seed plants could have "discovered" karrikins during the Cretaceous period, when flowering plants were rapidly evolving in the warm, oxygen-rich climate that led to frequent wildfires. Some plants became so dependent on karrikins that they are called "fire-followers" or "fire-ephemerals." Their seeds lie dormant on the ground until a fire burns, generating karrikins that bind to soil particles and then are washed by rainfall into the ground, creating "smoke water." The karrikins stimulate germination, and after those plants die, the next generation of seeds await, sometimes decades, for a new fire to generate karrikins and for the smoke water to wash over them.

What scientists recently began to discover about smoke, farmers have known for centuries. In *The Ecology of Plant-Derived Smoke*, the authors describe the practice of bathing fields in smoke among people in Germany, Britain, and Africa. In the seventeenth century, the French missionary Gabriel Sagard recorded a practice among the Huron people of suspending boxes of soil and pumpkin seeds above fires. In Guatemala, people burned the resin of *Bursera* trees, a relative of frankincense and myrrh, over their maize seeds.

After the crew's debriefing, I walked back to the trailhead with another volunteer. He was a retired lawyer, in his seventies. His banana suit and white leather gloves were smeared black. "How long have you been volunteering here?" I asked. He pointed to the fence posts along the trail we were walking on. "I put those up with my daughter when she was little," he said. Now she was an adult. He kept coming back to burn; he couldn't seem to stop.

After the reintroduction of fire, the population of Karner blue butterflies underwent an explosion. Each June, ecologists undertook butterfly population estimates, walking a transect through the lupines, carrying ten-foot poles and counting the number of butterflies. In 2020, they discovered a 500 percent increase over the year before. An estimated 46,100 butterflies were now in the pine bush, up from the mere few hundred decades before. The clouds of blue butterflies had returned.

I got home eighteen hours after I had left. Sitting at the table, I looked down at my legs, sliced with hundreds of tiny red cuts from brambles. It felt good to be tired, scratched to shit, filthy, and bathed in smoke. I was back burning at the bush a week later.

HENRY DAVID THOREAU once set a forest on fire. It was an accident, a campfire that caught on some dry grass. As he ran from the smoke and flames, he wrote in his journal later, he felt ashamed of what he had done. "But now I settled the matter with myself shortly. I said to myself, 'Who are these men who are said to be the owners of these woods, and how am I related to them? I have set fire to the forest, but I have done no wrong therein, and now it is as if the lightning had done it. These flames are but consuming their natural food.'" Having decided he was morally absolved, Thoreau stopped running and settled in for the show, which he described in his diary as a "glorious spectacle." The fire burned three hundred acres. Back in town, he was called a "damned rascal" and "woodsburner" for igniting what the townsfolk considered the last untouched woodland in the area. He called them "flibbertigibbets."

I had once asked Ben Wheeler, the wildlife biologist in Nebraska, "What was your first fire?"

We were driving down a two-lane road just outside Ord, oceans of grass on either side of us. A light rain splattered the windshield. He paused for a moment.

"Well, when I was a kid I lit my basement on fire and nearly burned the house down," he answered. "Twice."

I laughed.

"I had that universal fascination six-year-olds have with poking the logs in the fireplace," he explained. "The logs fell out. We had this red shag carpet, and there were two big burn marks on it for years."

"In the story of your life," I asked, "are those fires connected to your career in fire?"

Now he paused for a long while. "That would require some deep thought over several beers."

I was interested in why people started fires. Ecological benefit, as Dan Kelleher always emphasized, was primary. But I could see that wildland firefighters and prescribed-fire lighters had complex relationships with their chosen element. Respect. Caution. And, pleasure. They enjoyed being close to fire, manipulating it, talking about it, watching it. They were practitioners and disciples. The phrase "student of fire" is used so much in wildland firefighting that you can buy T-shirts emblazoned with it from thick supply catalogs that arrive in the mail each spring. It was first coined by Paul Gleason, a firefighter and mentor in the fire community. Gleason coined "LCES," which stands for "Lookout, Communication, Escape Routes, Safety Zones" and is a foundation of firefighter safety. Before he passed from cancer in 2003, Gleason had said, "I suppose I would want my legacy to be that firefighters begin to realize the importance of being a student of fire and that I was able to help make that happen."

On one level, what it means to be a student of fire is clear: to watch, observe, and study fire behavior. But the expression had taken on greater meaning over the years. It summarized a kind of posture of humility in relationship to fire and one's work, an acknowledgment that fire could never be entirely understood even if you spent a lifetime on the fireline. "It's setting your ego aside and admitting that you don't know everything," said Bre Orcasitas, a former hotshot. Did wildland firefighters love fire? I wondered.

Perhaps it was a tribe of would-be arsonists who had simply found a legal pathway for their obsession. In his book *On the Fireline*, the sociologist and former wildland firefighter Matthew Desmond painted a complicated picture of the subject. "The line between firefighter and arsonist is a fine one," he wrote. There were plenty of guys on Desmond's crew who admitted to having set fires in childhood. As one said, "I like to see it. It's amazing how something can go from nothing to just taking out huge trees. This one guy—we were in a fire investigations class, and he said, 'You know what firefighters are? A firefighter is a pyromaniac with his emotions under control.'"

I noticed an avoidance of the term "pyromania" among wildland firefighters. It was used in only the most frivolous way. Sometimes Dan would joke around before sending us out to burn. "What's that line from *Backdraft*?" he would bark, referring to the 1991 move about structure firefighters. "Burn it all!"

I looked up the scene later. Robert De Niro's character, Donald "Shadow" Rimgale, sits across from the pyromaniac Ronald Bartel in a prison visiting room.

"What about the world, Ronald, what would you like to do to the whole world?"

"Burn it all," says Ronald.

Years after he started a fire in the woods, Thoreau walked through the scar of an actual lightning fire on Mount Katahdin in Maine. The specific area was called the Burnt Lands. It was an open, meadow-like place with scattered blueberry bushes and few trees, a sight that seemed to send Thoreau into an existential spasm.

> Nature was here something savage and awful, though beautiful. I looked with awe at the ground I trod on, to see what the Powers had made there, the form and fashion and material of their

work. This was that Earth of which we have heard, made out of Chaos and Old Night.... It was a specimen of what God saw fit to make this world. What is it to be admitted to a museum, to see a myriad of particular things, compared with being shown some star's surface, some hard matter in its home! I stand in awe of my body, this matter to which I am bound has become so strange to me. I fear not spirits, ghosts, of which I am one,—that my body might,—but I fear bodies, I tremble to meet them. What is this Titan that has possession of me? Talk of mysteries!—Think of our life in nature,—daily to be shown matter, to come in contact with it,—rocks, trees, wind on our cheeks! The solid earth! the actual world! the common sense! Contact! Contact! Who are we? where are we?

I HAD NEVER felt irrepressible tendencies to start fires. But my mom always told me I was born with both a sun and moon in Aries, a fire sign, and that was the reason I made a lot of trouble. She also told me that when I was little, a seer told her that my group of friends, half a dozen girls all born within a few months of each other, had burned at the stake together in a past life. Later, I read that the first thing to burn when a person is on a stake is the eyelids, a morbid fact that stuck with me.

One day, I picked up a psychology book called *Pathological Firesetting*, published in 1951. The fire setter, according to the authors, psychiatrists at Columbia University, is a person overpowered by childlike impulses. They want to indulge in "the magical power of the child to create and destroy whole worlds." Most pyromaniacs, they claimed, were psychotics, psychopaths, mental defectives, imbeciles, morons, or "dull normals" (although 17 percent were identified as having superior intelligence), vagabonds, and gypsies. "No normal person sets fires habitually," they wrote. What I found interesting were the authors' descriptions of fire-setting as a desire that developed in puberty. Why, I wondered, was the desire to set fires associated with this period of change? In 1813, a doctor blamed fire-setting on disturbances in the brain during puberty because the eyes were deprived of arterial blood and people felt a need to create light. In 1873, a doctor described hysterical girls dreaming of fire during menstruation. Other doctors attributed the problem of fire-setting to nostalgia and melancholia. In the late nineteenth century, the French doctor Marandon de Montyel wrote that "pyromania often appears when the organism is born or dies from the sexual point of view—at puberty and the change of life."

Pyros were typically male, but the book contained around two hundred case studies of girls who set fires. Doctors believed pyromania afflicted girls between the ages of fourteen and twenty-five and that it was a self-destructive impulse, a kind of death wish and expression of suicidal desire. (The reason, they figured, that so many girls in rural settings were fire setters is because the city afforded more options and means for killing oneself.) Reading the case studies, I became convinced that, faced with similar circumstances, I, too, might have tried to burn down the world. Psychologists described the women as sexually unsatisfied psychotics who hated their mothers and resented their children as projections of themselves, but all I could see was how young, impoverished, and powerless they each were, broken by severe social restrictions imposed on them for their sex. Many of the girls and women had been exploited and abused.

In one case study a pregnant mother set fires around her children's bed after her husband deserted her. She said that her husband had tried to kill her by impregnating her many times. Another woman burned her home down after winning a "stork derby," a contest to see who could have the most babies in return for a monetary reward. A teenage servant girl was seduced and impregnated by a forty-nine-year-old man. Then her baby died. She began smoking a pipe and setting grass fires around his house every day. Another woman was interviewed in a state hospital. She had been committed once she'd received a diagnosis of paranoid schizophrenia after setting a fire.

"I object to a man who says, you are mine, do as I say," she said. "That is why I was angry with my husband. He abused my child and burned my mother's feet. I thought of shooting him, and I played with the idea of using a knife." She continued, "It's a man's world. They rule everything; perhaps they are smarter." Fire-setting, I began to think, was not a psychotic behavior but an expression of dissent by the otherwise powerless.

I became interested in historical cases of female arson, scouring the internet late at night for photos and old newspaper clippings. One that I found was an account of events that took place in North Carolina at the height of the Great Depression. In 1931, two halls of a reformatory school called Samarcand Manor were intentionally burned to the ground. Sixteen girls were arrested and charged as adults for arson, which at the time was a capital offense. As the historian Karin L. Zipf writes in her book *Bad Girls at Samarcand: Sexuality and Sterilization in a Southern Juvenile Reformatory*, the inmates at the school represented the quintessential disorderly woman. They were poor, uneducated, "white trash"—millworkers, union members, flappers, runaways. They were sexually promiscuous and rebellious. Ninety-one percent had been in jail and smoked tobacco (some of them were described as "snuff fiends"), and 70 percent had worked in the textile mills in horrendously exploitative conditions. Almost all arrived at the school with venereal diseases, and some were the victims of sexual assault, such as Margaret Abernathy, who was described as "bitter and cowed, having a droopy mouth and liquid brown eyes longing with hopelessness." Abernathy's father had raped her several times a week for three years. He was sent to prison, and Abernathy was sent to Samarcand.

The school was hardly a refuge for the girls. They were punished with switches, sticks, and razor strops, or their hair was shorn. Many tried to run away. Abernathy was one of the girls who lit a match on the evening of March 12 by going to the attic and setting a stocking on fire. "I thought that if I set it on fire they would send me out, and I was tired of that place," she said in her confession. The staff extinguished the fire, but another inmate, who was at Samarcand for "running around with boys," held a match to a paper dress in a closet and made sure the fire would burn. When the fires were finally out and two buildings lay in ashes, more girls confessed to the crime, hoping it would get them expelled.

In jail, awaiting their trial, the girls from Samarcand lit yet more fires. The second time, the newspapers reported, the girls had smashed all the windows and attacked with pocketknives the local firemen who came to extinguish the flames. The newspaper called them "fiends incarnate" and "fire bugs." Twelve of the girls were sent to prison. Within a few years of the fire, Samarcand began participating in a statewide effort to sterilize girls identified as mentally unfit. From 1929 to 1950, the state performed 2,538 forced sterilizations, half of which were on girls ages ten to nineteen; 293 took place at Samarcand.

I saw a picture of the girls. Taken by a photographer for the *News and Observer* on the day of the girls' sentencing, it was published on the paper's front page. It shows three of them walking up the courthouse steps in loose, sleeveless dresses typical of the twenties, their dark hair bobbed short. One girl's face is turned away from the camera lens; she is looking at her friend, who is holding a cigarette up to her mouth, her leg suspended above a stair, just about to take a step toward the courthouse. The photo caption reads, "One of the defendants is quite obviously taking the last final puff of her cigarette before facing the judge for sentence." The girls are mesmerizing, all legs, arms, cheekbones, and unrepentant youth.

AS A MEDICAL diagnosis, pyromania emerged in nineteenth-century Europe, a moment of intense political and economic turmoil. As a crime, arson had been feared by the ruling elite far earlier, as long ago as the fifteenth century, according to the historian Johannes Dillinger. Again and again, people in power warned the public of gangs of vagrants and street beggars—fire raisers—hired by subversives, political enemies, and revolutionaries to destroy towns and bring down political systems with arson. In the journal *Crime, History and Societies,* Dillinger recounted how the Habsburgs feared arson attacks from the Hussites in the 1420s, then the Venetians in the early 1500s, and finally the Turks during the next two hundred years. England feared arsonist Quakers, and blamed agricultural protests in the 1830s on Irishmen paid by Jews to set fires. Protestants in France feared Catholic arsonists. During the French Revolution, the rural population worried that the aristocracy was paying itinerants to set fire to their crops. These suspicions were almost always imaginary. Dillinger argued that the fact that arson was used in a variety of religious and political conflicts itself raises doubts about the reality of arsonist conspiracies.

Russia was the exception. Between 1860 and 1904, peasants, many of them women, lit so many fires that property adding up to almost 3 billion rubles was destroyed. Arson was seen as the primary obstacle to Russia's quest to become a modern society. Newspapers and official reports were full of vivid accounts of "fire mistresses," women who burned down the home or even the patriarch, according to historian Cathy Frierson. "Female arson was the embodiment of irrational fury," wrote Frierson. "Their willingness to use it against their neighbors, in-laws, or husbands was a warning that patriarchy writ large might be at risk in post-Emancipation Russia." As Frierson put it, who could doubt a woman's willingness to lay waste to the entirety of her immediate world?

The sociologist Michael Biggs has researched hundreds of cases of self-immolation over the twentieth century, beginning with the monk Thich Quang Duc, who lit himself on fire in 1963 to protest the Vietnam War. Nearly forty people, including several Americans, burned themselves in the years that followed in acts that spread beyond Vietnam. In 1969, self-immolations occurred in Poland, Czechoslovakia, Hungary, Scotland, and France. As many as a hundred South Koreans—mostly political leftists and union organizers—have burned themselves since the 1970s. Since 2010, it is estimated that hundreds of male and female Buddhist monks have self-immolated in Tibet and India. When the Tunisian street vendor Tarek el-Tayeb Mohamed Bouazizi lit himself on fire in 2010, it became the spark that ignited the Arab Spring.

I witnessed the aftermath of a self-immolation in 2018. It was when David Buckel, a sixty-year-old civil rights lawyer, lit himself on fire early one morning in Brooklyn's Prospect Park. Buckel was a practicing Buddhist, and the act was a self-immolation, perhaps the first, in protest of climate change. "My early death by fossil fuel reflects what we are doing to ourselves," he wrote in his suicide note. There was puzzlement over why Buckel chose the spot he did, just off the road in a patch of grass where it seemed few would witness him. But it was a place I walked by all the time, and I was there a few hours later with my four-year-old and a newborn, wondering why there were police cars. A friend who was an EMT showed me a picture of the body curled into a fetal position. Some speculated that Buckel had surrounded himself with soil so the flames wouldn't spread, creating black around him like Wag Dodge in reverse, but the singed spot in the grass where his body burned remained for months.

Perhaps it is too reductive to label self-immolation as protest. In 1965, the Buddhist monk Thich Nhat Hanh wrote a letter to the Reverend Dr. Martin Luther King Jr. explaining why monks like Duc chose to self-immolate, and why it couldn't be considered suicide or even protest.

> To burn oneself by fire is to prove that what one is saying is of the utmost importance. There is nothing more painful than burning oneself.... The importance is not to take one's life, but to burn. What he really aims at is the expression of his will and determination, not death. In the Buddhist belief, life is not confined to a period of 60 or 80 or 100 years; life is eternal. Life is not confined to this body: life is universal. To express will by burning oneself, therefore, is not to commit an act of destruction but to perform an act of construction, that is to suffer and to die for the sake of one's people.

IN THE ALBANY Pine Bush, I saw the disturbance of fire as something positive, a force for diversification and resilience that people could instigate. Until then, I had held an opinion of my own species as basically ruinous. Now I saw that we were not an unnatural presence on the landscape. People for hundreds of generations had assumed a responsibility to perpetuate the existence of this place. Fire's transmuting powers, the capacity to regenerate life through disturbance, captivated me. Drip torches and tools in hand, I felt like a link in a chain of mutual benefit connecting earth, grass, insects, animals, humans, and sky.

This was odd because sometimes the fire killed the very endangered butterflies we sought to conserve. They rarely travel farther than seven hundred feet in their lifetime, which typically lasts just four or five days. During our April burns, the spring brood of larvae were just emerging to feast on wild lupine leaves; by May the adults were fluttering about the bush in great masses. Unless they were on the very edges of the fire or it was low-ground fire, they probably would not survive it. A biologist on the crew warned us they would likely die. "We're OK with it!" she said cheerfully. Our burns created a mosaic of death and life. "God's work." Some butterflies would extinguish in fires this year so that next year, the ground would be crowded with wild lupine to feed a new generation.

4 MONSTROUS SERPENTS

IN AMERICA, THE LAND OF SMOKEY BEAR, SETTING FIRES REMAINS AN ACT LADEN with subversiveness. The roots of this contempt for fire on the landscape go all the way back to the European colonial era, according to historian Stephen J. Pyne, an emeritus professor at Arizona State University and a former wildland firefighter. Over decades of research on the global history of fire, Pyne has traced the first reference to the idea of fire suppression to India, where the British tried to implement a form of colonial forestry, simultaneously protecting forests from local use while also turning them into capitalist enterprises. In 1864, the British created the Indian Forestry Service and based it on French and German models of "scientific forestry," applying the principles of maximum sustainable yields and profits to the continent's tropical forests. Though Indian villages employed fire as a tool in their use of the forests, colonial administrators considered such practices irrational.

"Europe has a long tradition of distrusting and disliking fire. They thought it was wasteful and superstitious and disliked the smoke," Pyne told me. The British created a forestry cadet program in India, and graduates took the principles of scientific forestry to other colonial outposts—including Australia, Canada, and South and East Africa—where local fire-lighting traditions were inevitably quashed. (Though, as Pyne pointed out, this didn't work in Burma, where fire exclusion had a disastrous effect on the valuable teak forests and the locals revolted by engaging in labor strikes and incendiarism. "It's the only place in the world that I can find where fire suppression was officially renounced," he said.) That same year, at a forestry conference in India, an argument over whether fire control was possible or desirable broke out. On one side were the colonial authorities, military men and academics who saw the control of fire as critical for social control and protecting resources. On the other side were individuals on the ground, who proclaimed that not only was it impossible to control fire but that trying to do so would make things worse. Ultimately, the fire abolitionists won.

Pyne told me that the antifire rhetoric of the nineteenth century can be seen in terms of class. European elites were fine with fire in a hearth or a machine, but fire on the landscape indicated a lack of control. Setting fire was a kind of folk practice and the antithesis of Enlightened scientific thinking. "Woods burning had no more place in modern America than witch burning," Pyne wrote in his book *Tending Fire*. These attitudes were easy to import to America: fire-lighting provided another good reason to deride Native Americans and settlers, who as often as not had adopted fire-lighting practices. "I can find lots and lots of references to places in California of people burning because they picked it up from Native peoples," said Pyne. "Campers burned, ranchers burned, loggers burned, prospectors burned. There were just fires all over the place. The idea that settlers came in and immediately suppressed fire is nonsense. It was really the educated elites. They didn't have a science of fire at all."

A significant breakdown in historic fire regimes in the West, according to Pyne, was created by the overgrazing of sheep and cattle in the nineteenth century. "Bit by bit the old fire regime disintegrated, along with its practitioners," he wrote in *World Fire: The Culture of Fire on Earth*. "Intensive grazing removed the fuel necessary to propagate flame; roads and plowed fields interrupted sweeping firefronts." The first national park in the United States was established in 1872 at Yellowstone, and when the civilian superintendents failed to stop fires, the US Cavalry took over the task. In the summer of 1886, according to Pyne, troopers extinguished sixty fires. In 1905, when national forests were transferred to the newly created Forest Service, many of those landscapes' fire regimes were already trashed due to overgrazing.

The federal agency's first chief was a man named Gifford Pinchot. Pinchot had studied in Europe at the French National School of Forestry, and his goal was to apply the same rational scientific management to America's forests as had been achieved in India, which included producing high yields of board feet of timber. Pinchot, it turns out, was also a big fan of the eugenics movement and drew parallels between managing people and forests because "only in this way could the forest, like the race, live on." Strangely, Pinchot had traveled throughout the west and seen thriving, fire-adapted forests in person. He even understood and wrote about species like western larch, giant sequoia, pitch pine, and lodgepole pine and their adaptations to fire. Yet he wrote, "These facts do not imply any desirability in the fires which are now devastating the West." He still believed it was critical to snuff fire out of America's forests, especially after the 1910 "Big Blowup," in which three million acres burned across Montana, Idaho, and Washington in two days. The logic of total fire suppression became irrefutable.

It was not just foresters who thought fire was anathema to forest survival and prosperity. The founder of America's conservation movement, John Muir, held views similar to Pinchot's. Muir led the effort to create Yosemite National Park, a place that Native

American communities like the Ahwahnechee had burned for eons. But Muir saw fire—as well as "Chinamen," "Digger" Indians, Hispanos, herders, and sheep—as dirty and fundamentally destructive to nature, even sinister. Of the tribes who inhabited Yosemite Valley, he wrote that they were "mostly ugly, and some of them altogether hideous," and that they had "no right place in the landscape." It doesn't seem like a coincidence that he was friends with Henry Fairfield Osborn, who cofounded the American Eugenics Society. Muir wrote that fire destroys "not only the underbrush, but the young trees and seedlings on which the permanence of the forest depends; thus setting in motion a long train of evils."

When I spoke to Pyne at his home in Phoenix, amid shelves containing thousands of books collected over decades of scholarship, he pointed out how odd it was that Muir possessed this view of fire since he had grown up in Wisconsin, where prairie fires abounded. Muir's writings even contained close observations of fire on the landscape and descriptions of the way fire crept beneficially through Yosemite's groves of sequoias, clearing the ground for the trees' saplings to grow. "Muir understands fire, but he doesn't see it as necessary or useful," observed Pyne. "He thought it was something that you could remove from the landscape to make it better." Of course, in addition to being an environmentalist, philosopher, and botanist, Muir was a religious prophet. He believed that it was in wilderness, places pure and devoid of people, that one could experience a direct connection with God. Humans could never improve God's creation, only mar it.

FIRE SUPPRESSION DEMANDED putting fires out, but it helped to also demonize those who set them. Early antifire campaigns were aggressive, even violent. As the sociologist Kari Marie Norgaard described in her book *Salmon and Acorns Feed Our People,* at the very first meeting of the California State Legislature in 1850, legislators decided that "any person was subject to fine or punishment if they set the prairie on fire, or refused to 'use proper exertions to extinguish the fire.'" By deeming people arsonists for managing their land, suppression severed the connection between people and land. In 1916, a Native man by the name of Klamath River Jack sent a letter to the California Fish and Game Commission to explain the importance of intentional burning. "Indian burn every year just same, so keep all ground clean, no bark, no dead leaf, no old wood on ground, no old wood on brush, so no bug can stay to eat leaf and no worm can stay to eat berry and acorn. Not much on ground to make hot fire so never hurt big trees where fire burn. Now White Man never burn; he pass law to stop all fire in forest and wild pasture."

Two years later, a ranger on the Klamath National Forest wrote a letter to his supervisor proposing that missionaries be used to pressure people to stop igniting fires: "My scheme is as follows—Let the [Forest] service hire this woman to work amongst the indians on a general educational basis.... Her duties would be to travel up and down the river between Orleans and Elliots, stopping at different indian houses, talking to them in regards to their own welfare, but the principal point to impress on them would be the fire question. This woman can do more in one season towards causing the indians to adopt our theories in regards to fire than we can do in five."

The ranger also recommended that "arsonists" be shot. "Every time you catch one sneaking around in the brush like a coyote, take a shot at him."

In the early 1900s, however, some people saw the value of Native American fire practices and believed they should be adopted by the Forest Service. These proponents of intentional fire-setting were called light-burners, and they argued that by setting "cool fires" that burned the detritus and dead wood from the forest floor, humans could help prevent forests from becoming fire-prone thickets. One of these individuals was the forester Charles Ogle, who warned in 1920 that fire control was misguided and would only lead to greater fire hazards. "Nature has always taken care of the proper production of new growth and as the fires ran unchecked through the forests a proper amount of thinning was affected and the remaining trees were thereby given a better chance to mature," he wrote. The debate between light-burners and government agencies unfolded over years in newspapers and scientific journals. The Forest Service sneered at the idea of setting fires. According to Pyne, the agency called it "Paiute forestry" that was "beneath the dignity of an aspiring great power and Enlightened society." In some instances, Forest Service studies that inadvertently discovered the benefits of light-burning were stashed in file cabinets rather than published.

Meanwhile, in the southern United States, government foresters were having a hard time keeping people from burning in fire-dependent long-leaf pine forests. In response, the Southern Forestry Education Project was established, and from 1927 to 1930 it sent out teams to travel across the South, dissuading people from burning the woods. Called the Dixie Crusaders, the teams were modeled on roving revivalist meetings. They traveled some three hundred thousand miles handing out millions of pamphlets, playing motion pictures, and giving lectures. It was estimated that their message was delivered to some three million people.

Woods-burning remained enough of an endemic problem in the South, however, that the Forest Service eventually took a different strategy. The agency began hiring psychologists to go amidst the southern woods burners and study the reasons for their fire-setting ways. In one report published in 1940, a psychologist by the name of John P. Shea delivered his findings gathered from six months in the southern "piney woods" (aka the Blue Ridge Mountains) interviewing hundreds of people and asking them to explain their practices. "Woods burnin's right," one man told him. "We allus done it. Our pappies burned th' woods an' their pappies afore 'em. It war right fer them an' it's right fer us."

To the outsider, witnessing miles of free-running fires in the South was shocking, but the explanation, according to Shea, was that these were "poor whites" of average or above average intelligence but limited education. "The roots of the fire problem obviously go deep into the culture, the traditions and the customs of these people and their frustration," he wrote.

> The sight and sound and odor of burning woods provide excitement for a people who dwell in an environment of low stimulation and who quite naturally crave excitement. Fire gives them distinct emotional satisfactions which they strive to explain away by pseudo-economic reasons that spring from defensive beliefs. Their explanations that woods fires kill off snakes, boll weevil and serve other economic ends are something more than mere ignorance. They are the defensive beliefs of a disadvantaged culture group.

The author suggested a ten-point plan for diverting people's fire-lighting motives into "more constructive channels." (In his book about Florida fire history, Pyne wrote that there was likely no more a detested man in the southern fire community than Shea.)

That same year, a student thesis on the subject of incendiarism likened the tradition of woods-burning to "ancestor worship." In an effort to snuff out the practice, the author wrote that "the use of trained dogs to trail offenders has been quite successful. After one or two convictions as a result of using dogs, the number of set fires has in some cases been reduced to zero. Once the reputation of dogs for 'getting their man' is established, the appearance of the dogs with forest officers in an area is enough to stop the activities of the incendiarist."

LIGHT-BURNERS LOST THE debate in the early twentieth century. I sometimes wondered how things would have turned out if they had won. Instead, the logic of firefighting became unassailable for the next century. In 1935, the Forest Service chief, Gus Silcox, implemented what became known as the "10 a.m. policy," meaning that every fire start should be under control by ten the following morning. It was what Silcox called "an experiment on a continental scale." World War II created an even greater sense of urgency for suppression. Men were being sent overseas, and forests were understaffed and seen as vulnerable to fires and enemy attacks. In 1942, the Forest Service and the War Advertising Council created what was called the Cooperative Forest Fire Prevention Program, whose aim was to persuade the public to join the effort to prevent fires. One of the first characters to appear in the advertising campaigns was Disney's Bambi. Then someone came up with a cute bear as a mascot; he eventually became Smokey Bear. Other advertising posters used racialized imagery to purposefully associate forest fire with America's enemies. In the sociologist Jake Kosek's book *Understories*, he describes how America's forest landscapes had been turned into wilderness areas representing national and racial purity. The Forest Service's job, therefore, was to protect the forest from threats that "came from within, from those potential saboteurs, be they Japanese Americans or Communist sympathizers, who were working for the enemy." In this way, forest fires became part of a long tradition of using ideas of nature and wilderness as proxies for race and eugenics in American history, according to Kosek. "Nature and difference are held together by common social histories," he wrote. "Nature's repression, management, and improvement form well-worn paths that have defined the savage against the saved, the wild against the civilized, and the pure against the contaminated."

At the end of World War II, firefighting forces were inundated with military equipment and funding. The first hotshot crew was started in 1946. According to Pyne, the country entered a "Cold War on fire." Not that everyone embraced Smokey Bear. In some places, he was a hated symbol. One of those places was northern New Mexico, where heated social and political battles unfolded between federal agencies, environmentalists, and residents. In the 1990s, environmental organizations like the Sierra Club and Forest Guardians won injunctions to protect the spotted owls and what they described as some of the last small islands of "pure wilderness" left in New Mexico. Residents who depended on logging and firewood for their livelihoods were turned into outlaws by practicing a way of life that had been passed down for generations. In his detailed history of this period, Kosek described how Forest Service vehicles were torched, headquarters bombed, and environmentalists hung in effigy as symbols of new colonizers. Smokey Bear was called a "land thief" and "white racist pig." During this same period, the Sierra Club, the conservation organization founded by John Muir, proposed making anti-immigration one of its central policy issues, based on the belief that overpopulation from influxes of immigrants into the country was degrading the environment. The proposal lost, but some 40 percent of the club's members had voted for it.

The dendrochronologist and forest ecologist Tom Swetnam grew up in the contested landscapes of northern New Mexico. In 1950, the same year that the "real-life" Smokey Bear, a bear cub whose paws had been singed in a wildfire, was found in New Mexico, Swetnam's dad became the Jemez district ranger for the Forest Service. In the Swetnam home, firefighting was a fact of life: their house had a hand-cranked phone with a direct connection to a fire lookout tower. Swetnam collected old Smokey Bear posters as a side hobby throughout adulthood, eventually amassing one of the largest private collections in the country. "They are propaganda," Swetnam told me. "It's so inculcated in our culture that all fire is bad and that 'only you can prevent it.' It's stultifying, so many decades of the same message not really changing. Ultimately, it's like, come on, there is a different message that has to be embraced. Fire is also good. Can't we hold two opposing ideas at the same time? Isn't that the definition of intelligence?"

Throughout the twentieth century, the logic of wildfire suppression was that big fires could be put out by throwing more firefighters at them. It actually worked... for a while. Typically, 97 to 98 percent of wildfires were suppressed during initial attack. Some of those fires might escape initial attack and become extended-attack fires that lasted for a few days, and some of *those* fires might blow up into large fires, or what some wildland firefighters described as "big-ass fires." But even during the 1960s, there were clues that a dangerous paradox was being set in motion and that conditions for increasingly aggressive conflagrations were being created.

Typically, wildfires are variable in their intensity, scale, duration, and effect. Even a single fire might burn differently through a landscape depending on season, fuels, moisture, and topography, creating a mosaic pattern of severity that often benefited the diversity of plant and animal species. Different landscapes have different fire regimes. Some biomes, like boreal forests and temperate coniferous forests, might undergo a fire every century or so that burned hot enough to kill trees, a so-called stand-replacing fire. Grasslands might burn every year. Fire suppression imposed a single regime on all landscapes. In the western United States, for example, suppression led to a 300 percent reduction in burned acreage on average and long-term changes to vegetation. Significantly, forests across America became denser, and dead surface fuels accumulated.

Drought exacerbated the dangers of these continuous fuel beds by drying them out.

The seminal wildfire of Swetnam's youth took place in the summer of 1971. He was a teenager working at a gas station in Jemez Springs. Pumping fuel one day, he glanced up and saw what looked like a mushroom cloud over the mountain. His first thought was that Los Alamos, where the nation's nuclear laboratory was located, had been nuked by the Russians. But he quickly realized that Los Alamos was in the other direction. This mushroom cloud was smoke from a wildfire larger than any Swetnam had seen before.

The Cebollita Fire shocked the firefighters who worked on it. Over the course of a week, the fire ran through the crowns of trees. At one point it developed a fire tornado and blew a three-hundred-foot wall of flame across a canyon. One supervisor reported seeing flames a thousand feet high. Journalists covering the Cebollita described the fire in terms of a war zone with busloads of firefighters arriving every day and B-17 bombers flying overhead. The *Albuquerque Journal* reported, on June 7, "About 750 firefighters, many of them crack Indian teams, fought the fire through Sunday night. They were helped during the day by five converted World War II bombers dropping slurry and two helicopters dropping water." Other journalists described the fire as "unstoppable, destined to burn until it had destroyed every tree and building on the million-acre Santa Fe National Forest." Even the *New York Times* published a short piece on the fire burning thousands of miles away.

When the fire was contained after a week, Swetnam's dad accompanied an editor from the *Albuquerque Tribune* on a drive around the burn scar. The editor was shocked by the scene of still-smoldering root pits and torched trees. To him, the forest was dead beyond hope and might never return. "I was in Tokyo at the end of World War II," he wrote. "They were not talking about the A-bombs but about the great fire bomb attacks—particularly one night in

March 1945 when the 'fires of Hell' burned in Tokyo. They talked about the great 'fire storm' that swept the city. Ranger Tom Swetnam says that at its height, this great Jemez fire reached such fierce heat that it created a 'fire storm.' 'It was like a monstrous serpent writhing and twisting in the air,' he said."

The Cebollita Fire burned only five thousand acres, but it was an extraordinary event at the time. "No one had seen fires like that," Swetnam told me. "It just scared the crap out of these oldtime fire people."

Swetnam spent his scientific career gathering data that showed natural and human-caused ignitions in tree rings and advocating for reintroducing fire. "We can see the history of fire with the tree rings, we know these trees evolved with fire, we know what's happening with the lack of fire," Swetnam explained. "Most ecologists could see it. But the agency was still really reluctant to go there." This included his dad, who could never quite embrace his son's beliefs. "You don't really understand how crazy it can be," he would tell his son. "There are things science doesn't know about fire." And, Swetnam conceded, he was right. The Cebollita Fire was like a small omen of what was possible and what was to come.

IN AUGUST OF 1989, some scientists reported mysterious light detection and ranging (lidar) signals around Salt Lake City, Utah. Usually these kinds of signals indicated aerosols high in the stratosphere. At the time, the only event thought to be capable of generating enough convective force to push aerosols past the tropopause was a volcanic eruption. There was one other scenario, but it was purely hypothetical. In the 1980s, Carl Sagan and other scientists estimated that during a nuclear war, the fires ignited by the atomic blasts would be so huge they would create smoke plumes that pushed into the stratosphere and blocked the sun's rays, creating a nuclear winter. But in the summer of 1989 there were no active volcanoes (or nuclear blasts) near Salt Lake City.

After that, the same mysterious signals periodically reappeared on lidar scans. Confused, scientists began attributing them to unreported volcanic eruptions. Michael Fromm, a meteorologist and physicist with the US Naval Research Laboratory, began calling them "mystery clouds," and he started to wonder if they came from wildfire smoke. Then, on July 18, 1998, a University of Bonn lidar measured aerosols fourteen kilometers into the stratosphere over Sweden. Some hypothesized it was from a volcanic eruption on the Aleutian Islands, but Fromm and a team of scientists compared the observations with aerosol index data, lidar profiles, and forest fire statistics and determined that the mystery clouds were caused by a wildfire in Canada's boreal forest so big that the smoke plume had punched through the tropopause. "There were some who literally laughed when we tried to tell them what we thought was going on," Fromm told me. "If you saw aerosols in the stratosphere, it had to be a volcano." But by 2005, the evidence Fromm and his colleagues had gathered about these fire-generated clouds—Fromm has described them as "thunderstorms on steroids" or "fire-cloud chimneys"—was irrefutable. The official term for them, given by the World Meteorological Organization, was "cumulonimbus calvus flammagenitus," but Fromm and his colleagues named the phenomenon a pyrocumulonimbus or "pyroCb" for short.

For a little while, pyroCbs were thought to be black-swan events, extreme phenomena with little to no historical precedent. Then scientists started detecting them all over the world. On a single day in January 2003, wildfires in Canberra, Australia, created multiple pyroCbs. One of the pyro-convective cells generated so much energy it created Australia's first fire tornado, a category F2 with winds over 113 miles per hour. Western Russia reported its first pyroCb in 2010. Europe had its first in 2017, and South America and Africa each had its first in 2018. During Australia's Black Summer wildfires, there were thirty-eight pyroCbs. One of the blobs of smoke they generated grew to a thousand kilometers across, three times larger than anything ever measured before. As the blob hung in the atmosphere, it absorbed the sun's rays and became hotter, making it rise nearly twenty-two miles into the atmosphere, the same nuclear winter phenomenon predicted by Sagan.

Fromm told me we still don't have the historical data to understand whether these pyroCbs are truly unprecedented. Perhaps scientists have just gotten better at detecting them. But some consider them an indicator of a proliferation of extreme wildfire events around the globe. Rick McRae, a fire behavior analyst in Australia, defined extreme fires as those that release so much energy that they couple with the atmosphere. In his opinion, such events cannot be suppressed with traditional firefighting strategies. "When a fire develops a blow-up event, and couples with the atmosphere, it cannot be stopped, nor can its impacts be mitigated," he wrote in *Wildfire Magazine*. "The only realistic incident objective is to save lives, until the blow-up event abates." The global growth of extreme fire events—growth not just in the number of fires producing blow-up events but in the number of countries suddenly experiencing these phenomena—is, in McRae's opinion, "astonishing."

In the early 2000s, scientists began calling large, complex, and intense fires "megafires." Some were ignited by lightning, others by power lines or people. But all were man-made disasters: these fires had incubated in the landscape for half a century or more as fuels accumulated and the fire deficit grew. Megafires are big-ass fires transmogrified. They occur under extreme conditions: the hottest, driest, windiest days; often in dense, drought-stricken forests weakened by insect infestations or disease. Megafires could burn a hundred thousand acres in a matter of days, but it wasn't just their size that made them dangerous. They behaved in unpredictable ways that shocked and surprised seasoned hands and scientists alike. They burned at uncharacteristically high intensities and required huge suppression resources.

Global warming is exacerbating the conditions that lead to megafires. To learn more, I called the fire scientist Crystal Kolden, an assistant professor at the University of California, Merced. Kolden's demeanor was curt on the phone, and I assumed this was because it was fire season, a time when she is hounded by the media for quotes as each new wildfire explodes. But over the course of our hour-and-a-half conversation, I understood that her wariness was really a consequence of years spent trying to communicate two seemingly contradictory facts to journalists and the public. "We have this dichotomy of wanting to educate the public on the benefits of wildfire as a natural process," Kolden explained, "but also make clear that the types of wildfire we are seeing as a function of climate change—that is not natural at all. That is very much a function of what we humans have done by burning fossil fuels. It's a balancing act, and the nuance, of course, always gets lost in the twenty-four-hour news cycle."

On the day that I spoke with her, Kolden was tracking two pyroCb events: one over Mt. Shasta in northern California from the Lava Fire, and another from the Lytton Creek Fire northeast of Vancouver, Canada. The entire region, from central California to British Columbia, was undergoing a one-thousand-year weather event, a heat wave so intense that the highest temperature ever recorded in Canada, 121.3 degrees Fahrenheit, was recorded in the village of Lytton. The very next day, forty-four-mile-per-hour winds blew the fire into the village, killing two people and destroying 90 percent of the town.

"Is language an issue when we're trying to adequately describe what is happening now and what might happen in the future?" I asked Kolden.

"We do have problems with language," she answered. "When we look at what we're seeing now, we just have no historical analogue for that. We really struggle to convey just how extraordinary this is to be happening."

In 2015, Kolden had coauthored a study in *The International Journal of Wildland Fire* that used climate change projections to predict that the number of days conducive to megafires will increase by as much as 50 percent by the middle of the century. Projecting conditions, however, is different from understanding the wildfires they will spawn. "Scientists are some of the most informed people, and we understand full well what the projections are from the models," she told me. "Here's what these extremes will look like in the future; here's what we'll expect to see in area burns and days with extreme fire weather. Those are projections of the conditions. It is really difficult for us to actually try and model what those conditions will translate to in terms of actual fire behavior, activity, and growth."

One fire that shocked scientists was the Carr Fire in 2018 in northern California. "If I'm going to use the word 'unprecedented' for anything related to fire, that was it," said Kolden. The fire had started by accident: a tire blowout led to a vehicle's steel rim scraping on the pavement and generating sparks along a highway west of Redding. In a single night, it grew to twenty thousand acres. Three days later, the fire was burning so hot and generating such intense winds that a whirl of fire with winds of 143 miles per hour and reaching eighteen thousand feet high slammed into Redding. It picked up a truck and threw it a quarter of a mile down the road, killing the firefighter inside. It was the equivalent of a category 3 tornado. "I had never read a paper or ever heard a scientist even suggest that we could have something like an EF3 tornado formed because of a wildfire. For many of us who have worked on this for a long time, that was just really mind-blowing," said Kolden. "How did we have so much energy released on that fire under the right conditions that it was able to form a full tornado?"

Until recently, very little funding existed for fire scientists to study extreme fire behaviors, and a lack of data made it difficult to create useful predictive models. One reason for this is that it's difficult to collect data from a wildfire directly. But another reason is the legacy of fire suppression. When the goal was to extinguish every wildfire while it was small, researchers had little incentive to try to predict what fires would do when they were big. Until a couple of decades ago, Kolden explained, virtually no scientific study focused on combustion, embers, flames, or fluid dynamics in wildfires. "There's this legacy in the United States where our focus has been so much on suppression," she said. "The US Forest Service's basic message was: 'We're still going to fight these fires, we're still gonna put firefighters out there, so we'll just do the science specifically targeted towards making sure they don't die.'"

"Why do you think we didn't ask more questions about fire's potential?" I asked Kolden. Her response surprised me.

"The one thing that humans are uniquely able to do on earth is make fire," she said. "Fire is the thing that has been with us for so many millions of years, through our evolutionary history, that we basically have taken it for granted." Once humans started confining fire to boxes in our houses and automobiles and removed it from the landscape, we began to disdain and fear it. Kolden described how Native Americans, freed Blacks after the Civil War, and immigrant homesteaders were criminalized for setting fires to their land. "In my opinion, all of these social norms infiltrate into science," she continued. "We're not going to try to study fire, because we are a civilized society and we can control this. Why on earth would we want to waste time and energy trying to understand it?"

MEGAFIRES ACCOUNT FOR a growing proportion of the total area burned in America each year. Excising fire from the landscape was a century-long experiment that failed. "Imagine," Bre Orcasitas, the former hotshot, told me, "that we had the ability to control rain, and you just didn't let it rain for a hundred years. We've been controlling fire like that for a long time, just taking them out. Now we know that fire on the landscape is a positive thing; it's a necessary thing."

Megafires are challenging the very idea that suppression is possible. According to Timothy Ingalsbee, executive director of the organization FUSEE: Firefighters United for Safety, Ethics, and Ecology, firefighters have known for a long time that large fires defy suppression until weather conditions change or fuels run out. "Despite extensive scientific evidence that fire plays a vital role in maintaining ecological integrity in most western forests...," Ingalsbee wrote in the *International Journal of Wildland Fire*,

> managers now feel that they have grabbed a tiger by the tail and cannot let go—they are "trapped" by societal expectations to continue aggressively fighting nearly all unplanned ignitions because of the perceived high risks and hazards of letting fires burn. Thus, today's wildland fire research and management communities are increasingly experiencing the cognitive dissonance of knowing that ongoing attempts to exclude fire through aggressive suppression actions serve to increase the risks, costs and damages of fire over the long run, and run contrary to the community's land stewardship ideals and desires to restore fire ecology processes.

For me, it felt like a strange time to fall in love with fire. Huge swaths of the country were in the grips of historic drought. The dry period was biblical, so prolonged and parching that reservoirs were shrinking, groundwater disappearing, and wells drying up. Scientists were calling it a megadrought. Meanwhile, warmer air temperatures, less rain, and earlier spring snowmelt created prime conditions for wildfires. By June, the Southwest was predicted to be at above normal levels for significant wildfire, and the entire western United States was predicted to follow a month later. Wildland firefighters were bracing themselves for another grueling season following a record-breaking season when ten million acres burned. Lately, the fires themselves seemed to be changing, morphing into something impossibly dangerous, more wild and beastly than our collective imagination thought possible. What precedent—acres burned, homes lost, firefighters killed—would be broken this season?

I listened as Scott Stephens, a professor of fire science at University of California, Berkeley, described his studies in a place called the Blodgett Experimental Forest just west of Lake Tahoe. Fire scars in the trees show that, before 1900, fires burned every seven to fifteen years, a cycle that likely prevented large, high-severity fires because burned areas are unlikely to reburn within a decade. In 2003, Stephens ignited a controlled burn in a part of the experimental forest that hadn't burned in a hundred years. He returned several times over the years to light more controlled burns. Seventeen years later, the area had been transformed. The trees were healthy, sun reached the forest floors, and the dead and fallen surface fuels were sparse. "It's a forest that's much more capable of dealing with drought, climate change, and fire," Stephens said. "It gives us a sense that when you use prescribed burning in this frequent fire system, you're mimicking Indigenous fire and lightning fire that occurred for thousands of years."

Stephens had a nuanced take on the megafires that were afflicting the West. On maps, those fires were represented by red blobs illustrating their massive perimeter, giving the impression that everything within that perimeter had been torched. But a red blob can't accurately depict the effects of a wildfire at local scale. Some wildfires are doing needed restoration and bringing down the fire deficit. Some high-severity fires are even performing critical ecosystem functions in places that evolved with them. "If you look at the whole size of the August Complex," said Stephens, referring to the 2020 wildfires started by lightning strikes that burned over one million acres in California, "there's more restoration going on versus high-severity patches." After the North Complex burned a few hundred thousand acres in California's Plumas National Forest, Stephens flew over the area in an airplane with a group of congressional staffers and reporters. "We were looking at the high-severity patch that was really breathtaking," said Stephens. "But I was also looking out to the north side of the fire: it was all restorative."

This doesn't mean that megafires aren't devastating. According to Stephens, the historical range of high-severity effects of wildfires was somewhere between 1 and 5 percent. In recent decades, that figure has been 30, 40, even 50 percent, which means that tens of thousands of acres in a megafire could possibly trigger 75 percent tree mortality. "You're not going to get a forest back into a patch of high-severity fire of one hundred thousand acres," said Stephens. "The shrub seed bank will persist, but the tree seed bank has gone."

Stephens had been warning people of the potential worst-case scenarios of the fire deficit for twenty-five years. His predecessor at UC Berkeley was a revolutionary ecologist by the name of Harold Biswell who had delivered the same warnings before him. Following in the footsteps of the light-burners, Biswell advised returning fires to the West's landscapes back in the 1940s. He started working with California's ranchers to improve grazing, and managed to increase fire use in grasslands. Still, no one believed Biswell's claim that fire was beneficial for forests. When he started lighting fires on experimental plots in ponderosa pine stands, people were so upset that he almost lost his job. What Biswell discovered, however, was that the controlled fires didn't kill trees; they reduced manzanita and allowed pine regeneration. When a wildfire burned into his research plots in the 1960s, he saw firsthand that trees outside the perimeter of his prescribed-fire areas were damaged, while inside they were barely scorched. "You know, he was chastised for his entire career," said Stephens. "They just thought he was crazy to think that you could use fire proactively."

Sixty years had passed since Biswell argued that we needed to put fire back into forests, and now climate change was making the task trickier. A few days before I spoke to Stephens, he had participated in a prescribed burn not far from his study site. The day's weather forecast indicated that a low-pressure system would arrive to his north, far enough away that Stephens believed it would have no impact on the operation. As the day progressed, however, the humidity dropped and the wind shifted, forcing the experienced crew to start suppression operations to keep the fire inside the unit. "Frankly, for six hours we were in trouble. We had ten spot fires and slopovers all burning simultaneously," said Stephens. "We shouldn't have a problem burning in the Sierra Nevada at forty-three hundred feet in mid-April. That is nuts. There should still be heavy rain up there."

Stephens believed America needed a tenfold increase in prescribed burning, mechanical tree thinning, and managed wildfires—allowing lightning fires to burn naturally and only suppressing them under defined management strategies—in order to save and restore forests from the worst effects of climate change. "I'm optimistic, but I'm crazy worried," he said. "My view is that unless we move decisively in the next twenty years or so, forests are going to change right in front of our eyes. Some of these changes are probably headed our way anyway, because of climate change and also increased stress in different ecosystems. But the mixed conifer forest and some of these frequent-fire forests, I think we've got options. I just hope that we've got the wherewithal to begin that transformation. Because this is not the nineteen eighties or nineties. We're headed into a different realm."

5 EMISSARIES FROM A DIFFERENT REALM

IN HIS BOOK *YOUNG MEN AND FIRE*, NORMAN MACLEAN WROTE THAT THE ESSENTIAL ingredients to being a smokejumper were to be young and a little bit crazy. Smokejumpers also had an "it" inside them. "The 'it' is something special within," he wrote, "that demands we do something special, and 'it' could be within a lot of us."

Maclean's book is about the blowup at Mann Gulch that took the lives of thirteen smokejumpers. I read it two times. The first was years before I had ever worn a fire pack or held a drip torch, and the second time was after. It was like reading two different books—or maybe I was two different people. The first time I was reading about an esoteric world of Montana firefighting in 1949. The second time I was fully submerged in events I could visualize and almost smell: running up a slope, wondering what choices I might make if I found myself in a race with flame.

Maclean spent forty years researching the fire, visiting the steep gulch again and again, talking to the few survivors, driven by a desire to accompany the young men "whose lives I might have lived on their way to death." He had an unflinching ability to put himself in their shoes, even in the most tragic and painful last moments of their lives. "As the elite of young men, they felt more surely than most who are young that they are immortal," wrote Maclean. "So if we are to feel with them, we must feel they are set apart from the rest of the universe and safe from fires, all of which are expected to be put out by ten o'clock the morning after Smokejumpers are dropped on them." Their final thoughts, he figured, were not ones of fear, which burns away in the heat, but of self-compassion.

Maclean had reason to identify with the smokejumpers at Mann Gulch. He had worked for the Forest Service during World War I at the age of fourteen and fought several big fires in Montana. His brother-in-law fought on the Mann Gulch fire as a volunteer, and the two of them drove to the fire just days after it killed the smokejumpers. It had taken 450 firefighters to contain the forty-five-hundred-acre fire. Maclean wrote that he went to see it because he was possessed by his own black ghost, a memory of outrunning a fire as a teenager in a canyon. The fire had jumped across a trail, and Maclean tried to escape it by running uphill, spot fires igniting in front him, sprinting in switchbacks because his boots were old army boots rimmed with hobnails worn "smooth as skates." Burning alive on a mountainside was not one but three deaths, Maclean thought. First, the legs die. Then, the lungs. Then, the body in the main fire. When the young Maclean reached the end of his terrifying sprint, he had to put his shoelaces out because they were burning.

The rescue crews at Mann Gulch reported that the smokejumpers had repeatedly risen and tried to run after the fire reached them. Maclean wrote:

> The evidence, then, is that at the very end beyond thought and beyond fear and beyond even self-compassion and divine bewilderment there remains some firm intention to continue doing

forever and ever what we last hoped to on earth. By this final act they had come about as close as body and spirit can to establishing a unity of themselves with earth, fire, and perhaps the sky.

Michael Fromm, the meteorologist with the US Naval Research Laboratory who helped discover pyrocumulonimbus clouds, told me he thinks that the blowup on Mann Gulch produced a pyroCb, which may have contributed to the apocalyptic fire run that killed the smokejumpers. Maclean had spent years investigating different theories to try to explain the blowup, one of which hypothesized that a thunderhead had come in over the fire, generating winds that fanned the flames like a giant bellows. Fromm wondered if the fire itself had created the thunderhead.

In his book, Maclean described the conflagration at Mann Gulch as raging snakes, poisonous mushrooms, grey boiling brains sending up nuclear clouds, bulbous mushrooms impregnated by snakes, detonating ponderosa pines ("giant candles burning for the dead") whose disappearance in the explosion was "theological." "The world then was more than ever theological, and the nuclear was never far off," he wrote. He used the word "monster" over a dozen times.

Some wildland firefighters describe today's wildfires as dragons. You don't fight a dragon by standing directly in the path of its fire-breathing head. Instead, you grab it by its tail and then lop off its wings. Once it is weak and has lost its energy, you cut off its head. Wildland firefighters fight fires just like you would dragons—starting at the anchor (the tail), attacking its flanks (wings), and then, finally, chopping off the head fire. But the dragons have become increasingly unpredictable and extreme. PyroCbs are especially dangerous for wildland firefighters. The massive energy generated by the combustion of the fire always comes back down in the inevitable collapse of the pyrocumulonimbus formation. When it does, the energy takes the form of giant gusts of air called downdrafts. Downdrafts are now understood to be responsible for phenomena such as whirlwinds and fire tornadoes, creating embers that lead to rapid fire growth, cloud-to-ground lightning strikes, and erratic fire spread. Sometimes these storms generate phenomena that defy established wildfire-behavior science, such as the general principle that fires always move faster upslope than down.

I thought about the dragon-fighting analogy a lot. It conveyed something of the monstrous qualities of the wildfires themselves and hinted at the quixotic nature of fighting them. To be a dragon slayer in the twenty-first century is anachronistic, but won't we need to call on the same qualities—courage and strength—in the battle to limit climate change?

The anthropologist Anna Tsing has pointed out that Enlightenment-era Europe tried to banish monsters as irrational and archaic. But monsters have become useful again, Tsing argues. They illustrate how the world is constituted through the complicated entanglements of species and individuals and forces. Monsters are a way of thinking about "the wonders of symbiosis and the threats of ecological disruption."

I had taken to listening to a podcast called *The Anchor Point* on the long drive up and down Interstate 87 to burn at the Albany Pine Bush, on the advice of crewmates who said it would educate me about wildland firefighting. It was hosted by a former wildland firefighter, Brandon Dunham, who brought on men and women from across the country to talk about all aspects of fire from the philosophical to the mundane: boots, career paths, fire science. One evening I listened to an interview with a guy named Mike West, a former hotshot who had fought fires for seventeen seasons and had worked on some of the largest recorded wildfires in the country. He described near-death experiences and the resulting trauma. In

2014, after working the biggest wildfire of the year, in the Klamath region of California, he suffered a mental breakdown. "I remember we spiked out on the Happy Camp Complex for fourteen days, and it was pretty hard-core," he said. "I didn't see an automobile. I didn't hear music, very isolated." The day he got back from the fire, he sat with his fiancée and parents at a crowded restaurant, fighting back tears. "I'm having trouble being a person. I'm not there. I don't know what I'm doing. I'm freaking out," he recalled.

"You know that's a telltale sign of PTSD, right?" asked Dunham.

"Yeah, huge," Mike replied. "I just associated PTSD with Vietnam vets or combat vets. I was ignorant to the idea that it could be something going on with me. I just thought, 'I'm weird. I'm soft. I'm weak. Maybe I have an anxiety disorder that has nothing to do with fire.' ... I should have just screamed at the top of my lungs, put up the white flag. But I just couldn't do it."

His words made me consider that confronting a monster—what you saw out there, the feelings you experienced and things and people you lost—could exact a mental and spiritual cost. Mike West seemed like a kind of emissary, someone who had seen fire up close and was now warning us of what lay ahead. *Hic sunt dracones*—here be dragons.

BY THE TIME I contacted Mike, he had been out of the Forest Service for a year. He was now teaching middle school English and social studies in his hometown of Susanville, California. I wrote to his school email address, and he responded the next day. "I'm glad you're writing about PTSD and wildland firefighters," he wrote. "I'd be happy to talk to you for your story." Our conversations took place over weeks and then months, me pacing around or staring out the window, Mike sitting in his car in the early-morning hours to avoid waking his young children or, sometimes, because he needed to charge his phone and the electricity in his town had gone out due to a nearby wildfire.

I had warned Mike that talking with me might suck because I would ask probing questions and then fact-check everything all over again. I think he agreed to talk anyway because he hoped that our conversation might help people in similar situations be less afraid to seek help. Also, as I found out, Mike was just a really nice person: considerate and generous. And funny. One day he sent me two pictures: one of himself as a haggard, bearded hotshot in 2008, and one of him smiling brightly in a smart suit with a clean shave as a middle school teacher in 2021. "Here's a good before and after to warn people about life choices," he explained. "It definitely shows how a 25-year-old looks much older than a 38-year-old. Hotshotting with the reverse aging."

Mike had actually once dreamed of a career in stand-up comedy. As a wildland firefighter, he was the designated clown on crews, practicing jokes and sets on the line. But I noticed that Mike didn't actually laugh very much. I asked him about it one day, this complicated relationship he seemed to have with humor. Comedy, he told me, had been a cover for his anxiety for so many years that it eventually became a misery. He got caught in a psychological trap: if he wasn't funny, people would know something was wrong. He kept that fact hidden for a long time. There was a freedom for Mike in not needing to laugh or make anyone laugh anymore. He now chose to wear all his emotions, bad or good, on his sleeve.

Mike had grown up in Susanville with four brothers and sisters and parents who both worked at the local hospital. The town had a lot of wildland firefighters; the family's neighbor across the street was the superintendent of the Lassen Interagency Hotshot Crew, and his own dad had worked a season on a crew in the seventies. When he was five years old, Mike was hearing stories about the Yellowstone fires. Growing up, he thought fighting fire sounded cool but also scary. After graduation, Mike wanted to go to college to be a teacher, but one summer he applied to join a handcrew—a team of twenty people who suppress fire with hand tools—to make some tuition money. He was shorter than most but extremely fit, a high school football player who could do thirty-five pull-ups. Still, the work that summer hiking and digging line was the hardest physical labor he had ever done. Toward the end of the season, he was asked to serve a two-week assignment as a fill-in with the Lassen Hotshots. "There is this mystique to them," Mike said of hotshots. "My original opinion was, I didn't really like them. They seemed arrogant; I was kind of put off." He contemplated turning down the assignment but changed his mind. "I decided I was going to do it with the mindset of, like, 'OK, these guys think they're better than me. I'm gonna prove myself.'"

On that assignment Mike had his first close call. The hotshot crew went to Umpqua National Forest in Oregon and was cutting line midslope in a forest of ponderosa and lodgepole pine when some trees began torching below them. Mike was so new to fire he wasn't even sure if they were in trouble. "Are we in a bad spot?" he asked someone older than him. "Oh yeah," the guy replied. "We're in a bad spot." The crew ran down a ridge along the handline they had just cut, and the fire started ripping on either side of them. Mike recalled that the heat was intense and pine cones and limbs were falling on their path. He looked at the guy running in front of him, a friend from high school, and thought, Wow, we are going to die on this hill today. When the crew made it to their safety zone, Mike thought about his mom, as though the fear had turned him into a little kid. Afterward, no one talked about what had happened.

The season ended and Mike had mixed feelings about hotshotting. It was risky but also adrenaline-pumping work, a heady combination for a twenty-year-old. There was an adventure-seeking aspect to it that appealed to him. "Guys a few years older than me, they would describe these situations, they would show me pictures," he said. "They were talking about how much fun it was. I liked being with my friends and messing around, joking and traveling. That was one of the huge appeals to fire—that camaraderie and just having fun with your friends." Mike also realized why hotshots seemed arrogant. "The work they're doing is so insanely hard that you sort of separate yourself from the general public," he said. "Maybe you think you're kind of special or something."

The superintendent of the hotshot crew told Mike he had done a great job and encouraged him to apply for the next season. So, he did. "In my mind, I kind of morphed into it," he said.

ONE HEARS THE word "wildfire" and imagines a wave of flames moving through a forest. But wildfires are amorphous and dynamic, shape-shifters determined by changes in wind, topography, temperature, and humidity. Parts of a wildfire are flame, while other parts are ember. It can hush and then roar and hush again. Flanks can grow into new fronts, or fronts can reverse themselves and reburn areas. When Mike thought about the close call he'd had during his first full season as a hotshot, he remembered how quiet the fire had seemed and how it had meandered around the duff on the ground, barely smoking.

It was late June and lightning strikes had ignited fires in the Sky Islands, isolated mountain peaks surrounded by the ocean of the Sonoran Desert in Arizona's Coronado National Forest. The Nuttall Complex, as it was called, was several thousand acres by the time Mike and his crew were working on it, but it wasn't anything crazy, just skunking around. Quiet. The crew camped at eighty-eight hundred feet. They dug line in the mineral dirt on a ridge and then lit backfires to starve the main fire of fuel.

At seven a.m. the crew had a morning briefing. During the night, some of the flames from their backfiring had burned over their line. The slopover was on the east side of a narrow ridge along a canyon. Mike and his crew began hiking into "the hole"—the area downhill from their escape route—to dig line around the slopover. There were three other hotshot crews working that day. It was rough terrain, rocky and steep. At certain parts, the slope dropped away at seventy-five degrees. Mike carried a forty-pound pack and a Super Pulaski, a tool with a steel axe blade connected to a thick grubber for digging in hard soil. At some point, the guy hiking behind him said, "Hey, West, where are your sigs?" Mike realized he had forgotten the liter-sized metal bottles containing chainsaw gas and bar oil that the saw teams needed to operate the chainsaws. The mandated punishment for this mistake would be two hundred pushups for each bottle. It was too late to go back and get them, so he tied a sweatshirt over his pack to hide the mistake and planned to fess up back at camp that night.

The Lassen Hotshots worked the southern edge of the slopover near another crew, the Plumas Hotshots. Mike was digging line next to a buddy who had more experience and gave him pointers. "You need to move this rock," he would say. Or, "Your berm is too high." The hours passed by, and when they sat down to eat lunch, Mike was starving. He ate an MRE sitting on the ground when his friend turned to him.

"W, look," he said. "Look at that smoke column." Mike glanced up and saw a churning black mass in the sky.

"I don't think we should hang out down here much longer," said his friend.

There were seven lookouts on the Nuttall Complex. One of them was the assistant superintendent for the Flagstaff Hotshots. He stood on a rocky knob further down the ridge from where Mike was working. Below the lookout, down in the canyon, was the

Nuttall Creek, and the lookout watched the opposite ridge, over which the main fire was burning. Around eleven a.m. a weather update came in over the radio. "Humidities will be bottoming out in the next couple hours," the incident meteorologist said. The Haines Index, a measurement of the instability and dryness of the air on a scale of one to six, was going to head into a "super six." "Everything looked real benign, you know; it was real light smoke," the lookout later recalled. "There wasn't anything real heavy going on." At around 12:30 p.m., the lookout was still watching the ridge, but he swiveled his head to check on the line and then turned back. The whole movement took about five seconds, but now he saw a crown fire, three hundred yards wide, roaring down the slope toward the creek below him. It was a blowup: a sudden increase in fire intensity accompanied by violent convection. "I never saw anything ever that indicated to me that was getting ready to happen," said the lookout. "There's five hotshot [superintendents] with over two hundred years of experience, and no one saw it coming."

One of the first principles of fire behavior that every rookie learns in training is that fires move uphill faster than they do downhill. Hot air rises and preheats the fuels above it, making them drier and readily available for combustion. The steeper the slope, the more preheating of fuels, the faster the rate of spread. For every ten degrees of uphill slope, a fire will double its speed. The spread of fire is measured in chains per hour; a single chain is sixty-six feet. An uphill fire might move at several hundred chains per hour, but for every ten degrees of slope in a downhill fire, its speed is cut in half. That was not what the lookout was seeing. This was a downhill-running crown fire, in which the energy of the fire was incinerating the crowns of trees. Once it reached the bottom of the drainage, it would sprint up the slope to the ridge the crews were working on. The lookout radioed the superintendent of the Flagstaff Hotshots, Paul Musser, who was located further down the ridge.

"Hey, Paul, I can see fire on our side."

"Yeah, it's down below us," said Musser.

The crews' leaders told everyone to get ready to leave. "Get up, start lining up, we need to go." As Mike gathered his things, the minutes seemed to elongate. The Plumas Hotshots realized a crew member had left to go to the bathroom in the woods and began a frantic search. "I was standing in line just thinking, 'Can we go? Let's go let's go let's go,'" recalled Mike. Impatience became tinged with fear. By the time they started hiking out of the hole, the fire was beginning to boil up the slope to their right. They moved as fast as possible, sometimes crawling on their hands and knees over boulders, grabbing each other's hands and tools and pulling one another up steep rock faces. Crew members started to fall behind, and the line started to gap. Mike's squad leader began yelling. "Just go, just go. Pass 'em!"

Mike was struggling to breathe with the exertion and wondered if they would have to drop their packs to outrace the fire. The crew passed the chainsaws back and forth after the sawyers got too tired carrying them. Mike had never seen his buddy lose his cool before, but now he was shouting, "This is some serious shit—let's go!" By now, the sound of the fire was overwhelming. It was a detonation with no end. The roar of a running crown fire is sometimes compared to that of a freight train or jet engine, the thunder of heavy ocean surf. Mike told me it reminded him of the explosive noise of a waterfall. He could see that between him and the fire lay a stretch of unburned fuels, what firefighters call "the green." "The fear is that you have this active, loud, roaring fire and you're in the green and you have nowhere to go and it's just, is this going to get me?" said Mike. "The monster is coming. What's it going to do?"

The crown fire raced uphill, torching the trees to their right. They watched as the flames reached the ridge in front of them, then flicked over the edge and stopped. They pushed past and finally got to the top of the mountain. It had taken the group of around forty hotshots half an hour to get out of the hole; at the top, Mike threw up. Then he realized that not all four crews had made it out, including his own superintendent.

A weird feeling set in as they waited around a small lake. Some people fell asleep. Others joked and ate snacks. Some just stared silently for long periods or walked into the woods to be alone. Mike watched Skycrane helicopters sucking up water from the lake and dropping it on the fire. Behind the choppers, the smoke had risen so high that it had combined with the atmosphere to become a pyroCb cloud—a dark column of ash, smoke, and water vapor twenty thousand feet tall. At the top of the formation, where the ambient air temperature dropped below freezing, the vapor froze to create a smooth white surface, like a meringue—a process called ice capping. It was the first time Mike had seen it.

Over the radios, the firefighters heard that eleven hotshots in the hole were deploying their fire shelter.

"This is really bad," Mike said to his squad leader.

"Yep," the man replied tersely.

"When we were hiking out, I thought, Is this going to be a fatality event where I die?" Mike told me. "I was super scared. But we got out. Then I hear shelter deployment, and I'm thinking, How many people are going to die down there?"

WHILE MIKE WAS digging line on the slopover's southern edge, a wildland firefighter by the name of Thomas Taylor, known as "Thom," was cutting trees and brush on its northern edge. He was a sawyer with the Flagstaff Hotshots, and he described the work that morning as feeling normal, familiar. "Cut as fast as you can and try to make good decisions, drink plenty of water and, you know, work hard and have fun," Thom later told an interviewer. Around noon,

Emissaries from a Different Realm

Thom became aware of some radio traffic between his crew and the lookout up on the rocky knob. The lookout said something like, "The main fire's not an issue." At that very moment, someone walked by. He stopped, looked Thom in the eye, and said, "The main fire's *always* an issue." Then he kept walking. It was a premonition, but he didn't know it yet.

Thom and his saw partner soon headed further down the ridge to a flat area flanked by steep rock chutes and scattered pine trees. It had been used as a helicopter landing spot earlier that day and was called H4. Their assignment was to prepare the area for backfiring by cutting snags and trees and throwing them over the edge of the ridge. As they got to work, the fire started to make its downhill run, and their superintendent, Musser, arrived to get a better view of the fire's spread. It had now crossed the Nuttall Creek and was burning in the canyon below them.

Thom and the rest of the Flagstaff Hotshots were in a bad situation. Their escape route was the same hike up the ridge that the Lassen Hotshots and other crews were now undertaking. But they were way deeper down in the hole. The crew's superintendent saw the fast-moving fire and canceled the operation, telling everyone to get to H4. The crews began running down the ridge when the lookout saw the flames reach up over the edge and curl over the head of the person in front of him. "Hey, we just got cut off by fire," he told the superintendent over the radio. He looked around and remembered thinking, "Man, we're standing right in the middle of the saddle. It's going to hit us right here. This is going to be it."

The only choice left was to retreat to a nearby stand of aspen trees, which grow in moist places and are therefore less flammable. It was the closest thing to a safety zone they had, and the lookout told people to drop their gear to get there, if they needed to. As the lookout waded through shoulder-high brush, he kept glancing over his shoulder to keep an eye on the roaring fire. He stumbled on a firefighter lying in the dirt, staring up at the sky. The lookout didn't know it at the time, but the man had rhabdomyolisis: his muscle tissues were breaking down from exertion and poisoning his bloodstream, kidneys, and heart.

"Hey! What are you doing?" the lookout yelled. "You got to get up! We got to get out of here; this isn't going to be a good spot." The firefighter didn't respond, so he yelled some more. "Hey, man, you need to get up *now*, now's the time, we got to get up and go, can't stay here!" He took the firefighter's pack off and began carrying him to the aspen stand when he reached a rocky section that he knew he couldn't negotiate. As he put the firefighter back on the ground and continued to try to revive him, three hotshots appeared through the smoke.

"Hey, do you need a hand?" they asked.

"Yeah, I need a hand!" the lookout said.

Back at H4, a shower of embers began to ignite the duff around Thom as he worked. The smoke became so thick that it blocked out the sun. Thom teased his saw partner, who was now urgently digging around his pack trying to locate a flame-resistant shroud to protect his neck from the heat. Then a three-hundred-foot wall of flames came out of the canyon and over Thom's head. A pilot circling in a helicopter above H4 described seeing a solid-black column of smoke rise out of the canyon and begin to twist itself into a fire whirl, a vortex of hot air and superheated gases. The whirl carried smoke, debris, and flames, and as it spun up the slope and over the ridge, it started to bend, laying itself horizontally as it continued to rotate. There were ten other firefighters along with Thom on H4, and the wind from the fire whirl blew someone's helmet off their head.

"We started bucking stuff up and throwing shit over the hill," said Thom. "But once it started making its runs and coming up out of the hole constantly, you know that's when I started freaking out, basically."

Thom and his partner realized that nothing they did was going to prevent the fire from approaching. They were stuck on H4. The incident report contains a photograph of Thom at this moment. He is a dark silhouette on his knees, his head bent downward, his shoulders hunched against the flames behind him. Thom said it is difficult for him to look at the photo because it captures the moment when he felt like his brain was eating itself with fear and he began to hyperventilate. His saw partner tried to calm him, and another firefighter dumped out a lunch and gave him the paper bag to breathe into. Thom's breathing regulated just enough to help him recognize that he needed to act. He started digging into the ground and preparing to deploy his shelter.

I ONCE HEARD a wildland firefighter describe deploying their shelter as stepping into their own coffin. Fire shelters are the last available option for situations with no escape. American wildland firefighters are the only firefighters in the world who carry shelters, and they also have the highest rate of burnovers. In every other country, firefighters are never expected to get so close to a fire that it could burn them over. To maintain qualifications as a wildland firefighter, everyone annually practices pulling a folded test shelter out of its plastic sheath, unwrapping its protective cover, shaking its folds apart, climbing into its innards, and falling to the ground face down as fast as possible, being careful to create a seal around its edges to keep out superheated gases that will burn your airways and kill faster than flames can. Sometimes practice shelter deployments are done in front of giant fans to mimic the winds of a wildfire bearing down on you and threatening to rip the shelter out of your hands.

Burnovers can happen so fast that people's skin can start to cook before they get into a shelter. One wildland firefighter in South Dakota described putting his hands up to protect his face and airways and feeling like there was water running down his hands. It was his skin coming off. "At that point in time, it was weird," he said. "There's this surge of adrenaline. I think it's your mind telling you, 'Oh, shit, you're burning.' I had this immense power in my legs telling me it's time to run.... But my mind's telling me, 'You're not running, you're staying put.'"

Once you get in a fire shelter, you don't know how long it will be before you can get out. It could be minutes or hours. Rookies and veterans alike often retell this part of their experience through tears. "It's hard to stay in that shelter because they get so hot," said one person. "You just got to keep thinking it's a hundred times hotter if you jump out." In the dark, smoky shelter with the heat bearing down, the urge to look outside to see if you should run is almost insuppressible. People describe thinking of their family and friends, the possibility that this is their last stand, and making their peace.

"It's just real scary," said a Zuni Hotshot who was forced to deploy during a wind shift in the Holloway Fire, a grass and sagebrush fire near the Nevada-Oregon state line in 2012. "The only thing I could think about to help me with my breathing was to dig a little hole in the ground." She survived one flame front and then another and another and another before her crew could rescue her. Other wildland firefighters recall coming out of their shelters and not recognizing crewmates because their faces were covered in dirt from trying to breathe.

Thom Taylor had deployed a shelter once before. Three years earlier, he had been a squad boss for a handcrew on the Okanogan-Wenatchee National Forest in Washington. They were sent on an initial attack of a small fire in the Chewuch River Canyon started by a campfire. It had been an unusual winter—the second-driest in thirty years—and on that July day, the temperature reached ninety-four degrees Fahrenheit. The Energy Release Component, a measure of potential fire intensity, was at its highest historical maximum since officials began keeping records in 1970. Thom's crew put out spot fires all day, some working with only a few hours of sleep, and ate lunch on the dirt road that snaked alongside the river.

That afternoon, what became known as the Thirtymile Fire began to dramatically transform, doubling in size from fifty to a hundred acres in just fifteen minutes. At 4:03 p.m., someone at a nearby lookout tower reported that the fire was developing its own thunderhead, an indication of a pyroCb. Thirty minutes later, the flames had jumped across the road leading out of the canyon and blocked the crew's escape route. The crew drove deeper into the canyon, fourteen people in a single van, not realizing that the road was a dead end. They found a place next to the road and stopped, Thom puffed on a cigarette before climbing up onto a rock scree to get a better perspective of the fire. He could see two separate smoke columns running up both sides of the canyon like a V and exploding. The fifty-mile-per-hour winds sent a firestorm of ash and embers ahead of the flaming front.

Thom tried to run back down to the road but hit a wall of superheated gases. He retreated to the rocks and pulled out his shelter, letting the wind open it like a parachute. Moments before he stepped into the shelter, he saw a group of five people running up the rocks toward him as the flames chased them. He yelled, "Deploy! Deploy!" Inside his shelter, Thom was overcome with terror. He sat in a fetal position and tried to breathe as the heat and flames and winds battered the rock scree. He could hear people praying, talking, and screaming from inside their shelters nearby. The upper left corner of his shelter began to burn and fill with smoke. He tried to think. "If I stay in my shelter I will die. If I choose to leave my shelter, I will die. But I can't give up." He fled down the rocks, running through the flames to the road, and jumped in the Chewuch River.

Of the five people who deployed next to Thom on the rocks, four died in their shelters of asphyxia from inhaling superheated products of combustion. The heat had been so intense that the boulders near Thom's shelter had fractured. "Trust the feelings you have. Listen to your gut," he said in an interview afterward.

Thom joined the Flagstaff Hotshots the next year. Now he found himself on H4, deploying his shelter once again. He remembered shouting apologies to his superintendent from inside. He was overwhelmed by a feeling that he had fallen apart emotionally and let the crew down. When the heat and embers lessened, Thom and the other firefighters emerged from the shelters. Thom took a picture of his melted chainsaw. Once deployed, shelters are meant to stay at the site so investigators can return and analyze the events, but Thom thought, Screw that, I'm taking this with me. He carried the shelter out of the hole with him in case he needed to deploy again.

AFTER THE THIRTYMILE Fire, Thom's coping mechanism was to finish work, drink a fifth of Crown Royal whisky, pass out, wet his bed, shower, and start the whole process over again. He quit the Flagstaff Hotshots after the Nuttall Fire. He felt he needed a job where he was more in control, so that August he went to mark timber in the Wenatchee National Forest. Two weeks into the job, he turned on the radio and listened as a nearby helicopter's tail rotor hit a snag and the copter smashed into a helispot, killing the pilot. Earlier in the season, Thom had flown with the same pilot in the same ship. His first thought was, Man, I can't get away from this. His brain started to eat itself again.

Thom experienced nightmares, especially when he was sleeping in a tent. He would sometimes cut his way out of the tent with a knife. He suffered from intense episodes of anxiety that manifested as what he described as heart rolls, in which the organ seemed to release a warning to his brain that he was about to have a heart attack and die. It was hard to predict when they might occur, but quiet moments were dangerous. Looking at a fire scar as a fog rolled. Staring out over Bellingham Bay. Rain. Long drives. One afternoon his heart rolled while he was riding a chairlift up a mountain in Idaho and it was "snowing chicken feathers, no wind and deathly quiet. Just as it is after a shelter deployment." A few years after the Nuttall Fire, during one emergency room visit, Thom's heart rate was measured at three hundred beats per minute.

Mike West also struggled after the Nuttall Fire, but he didn't tell anyone. The day after the crew's escape out of the hole, he went back to work, pushing himself so hard his squad leader intervened, dumping canteens of water over his head. Mike learned to work at a pace just below what would trigger sickness or heat exhaustion. He became a sawyer on the crew. His friendship with his saw partners, the "swampers" responsible for clearing the brush he cut, were so close it was almost like being in a marriage. They screamed at each other over the noise of the chainsaws. They spent so much time together that he knew their habits and movements before they did, how they would place their hands on a branch. He came back to the crew, year after year. "Sawyering took me to places in my mind that I'd never gone before," said Mike. "It was almost like chasing a high."

Wildland firefighters talk a lot about "embracing the suck." Crews often go through hell together. "Everything is earned and learned through pain, hard work and teamwork," wrote Aaron Humphries, the former superintendent of the Eldorado Hotshots.

"The relationships with crew members are indescribable. The moment you realize that the team is more important than the individual and everyone brings forward their best to help the team. That on any given day someone will need you to pick them up because today is not their day. Knowing your turn is coming to be the one in need of being picked up." Another hotshot, Chris Mariano, wrote:

> A true commitment to being a Hotshot is born out of love for the job and service to our landscapes and communities. The sense of purpose is unrivaled by anything else I know. The required physical and mental fortitude are extreme, but the bonds created within a Hotshot crew are unparalleled. Many of the best moments of my life came when I was completely depleted; hungry, tired, beyond mental and physical exhaustion and yet I was completely fulfilled. Heart, grit, camaraderie, sense of duty and purpose is what sustains individuals through a season which is punctuated by hardship and tragedy.

Despite the hardships of the job, wildland firefighting has long had a culture of stigmatizing talk about mental health. Melissa Peterson is a former wildland firefighter who works as a licensed marriage and family therapist in Reno, Nevada; she met her husband fighting a wildfire and treats wildland firefighters at her practice. "In wildland firefighting, there are these requirements and expectations to be tough and self-sufficient, not having emotion, not showing weakness, not asking for help, not saying that you're struggling," Petersen told me. "That is for a purpose. Because it is really hard to fight fire if you are overwhelmed by a fear of fire." Peterson read aloud from the criteria for PTSD as outlined in the *Diagnostic and Statistical Manual of Mental Disorders*, which includes exposure to actual or threatened death or serious injury. "Just by doing their job, they are in a position to meet the first criteria for PTSD."

Emissaries from a Different Realm

The flip side of Mike's high as a sawyer was intense fear. "I remember certain times cutting line and in my mind being scared, thinking, Are we going to have to run today?" he said. "I spent hours and hours and hours with knots in my stomach." Mike's fear was that he would burn to death. "There are all these other things that could happen, of course, on the fireline—vehicle accidents, aircraft and trees falling," he said. "But those things didn't bother me as much as this fear of getting burned up." He didn't talk about being fire-scared with anyone.

In the off-season, Mike would go on unemployment, like many seasonal federal employees, or enroll in college courses. He often rented a house with other wildland firefighters, and they would snowboard and drink together. In the winter of 2009, Mike moved to a little mountain town in Northern California with his childhood friend Luke Sheehy, a smokejumper. Every day with Sheehy was an escapade. They woke before sunrise to jog through miles of snow and lift weights at the local high school's gym. They played practical jokes on friends who crashed on their couch, and they took road trips together. Mike worked on his stand-up material. He started spending time with a beautiful, blonde-haired teacher from Susanville, Cassie Dunn.

Before every season started, Mike would say a sort of prayer. "I accept that people are going to die this year on the fireline," he would say. "I hate it. I really, really hope it's not me or anybody that I know." He knew it was a numbers game. The longer he stayed in fire, the more people he knew. Mike's list of close calls grew. Sometimes he forgot his way around Susanville, a town he had lived in since he was a young child. "I felt like I was losing my mind," he said. In the summer of 2013, Mike dislocated his shoulder at work. A few days later he got a phone call at home. Sheehy had been killed when he parachuted into California's Modoc National Forest to put out a lightning-ignited fire on a single white fir tree. He was struck by a falling limb and died before he reached the hospital. Sheehy was twenty-eight years old. "I just remember this scream coming from [Mike]," said Cassie. "And he couldn't talk, and I was just holding him. I couldn't fix it. Luke was a big part of our lives. So, yeah..." A couple of weeks after Luke's funeral, the Yarnell Hill Fire in Arizona killed nineteen crew members of the Granite Mountain Hotshots in their fire shelters when they were overrun by a blowup that caused a pyroCb. "I remember being very upset about it, but I didn't really process it or think about it," said Mike. "I was still in such shock from Luke's death."

For Cassie, the pressures of Mike's job were becoming unbearable, especially after their son and daughter were born. "He would go to work in the morning, and he could come back that night or in two weeks or four weeks," she said. Mike asked her not to talk to anyone about the problems he was having, the nightmares and fear and anxiety. But one day she pleaded with him. "You have to acknowledge that this is affecting our lives," she said. "You don't have to apologize for something you can't control, but I need you to acknowledge that our lives revolve around this every day." She started talking to other women who were in relationships with wildland firefighters. "Is this happening in your life?" she asked. She discovered that many of them were on antidepressants to cope with the stress of their partners' jobs.

In 2017, Mike was working as a fire prevention officer on the Plumas National Forest, and a small arson fire escaped his control. It grew to sixty acres. The following day, as an engine fought the fire, a pump malfunctioned, causing a hose to spray boiling water over a firefighter. "I felt like, if I had gotten that fire out, this guy never would have been burned," he said. Around that time, Mike began thinking that killing himself was a good idea. It was a thought that floated around and comforted him when he felt trapped by not having enough money, by not being a good family man, by hating a job that he used to love. Mike wasn't alone. In 2015 and 2016, fifty-two wildland firefighters died by suicide in the US—twenty-five more than were killed in the line of duty.

ONE DAY MIKE was in his office and checked the website of the Lessons Learned Center, a federally funded organization tasked with improving safety for wildland firefighters through education. He noticed that they had a podcast episode about the Nuttall Fire, and he started to listen. It was an interview with Thom Taylor. Mike didn't know Thom, or even that he was one of the firefighters deploying down on H4. Now he listened as Thom described his experience publicly for the first time, being hunched over with the monster at his back, his brain cannibalizing itself. Thirteen years later, the memory caused Thom to break down in tears during the interview. "[Nuttall was] the first time since Thirtymile that I've been in a situation where, you know, it got hot. And I, you know, missed opportunities to make better decisions because I wasn't able to focus. Because I freaked out." Thom's voice trembled. He paused for a while. "That's why Nuttall, for me, is more difficult than Thirtymile. Because I was embarrassed by my actions."

Thom talked about his long road to understanding trauma, about talk therapy and Xanax, of learning how the brain reacts with the body, and about the culture of wildland firefighting. In his whole career, he said, he had never gotten a call from his bosses to check on him, to say, "Hey man, how's it going? What's up? You doing all right?"

In a blog post for the Lessons Learned Center, Thom questioned the adequacy of LCES, the acronym designed to teach wildland firefighters how to implement safe procedures and identify hazards on operations. "Is LCES Dead?" he asked. The simple acronym, with its instructions to "Have a communication plan" and

"Be in a safe location," failed, in his opinion, to capture the complex emotional, psychological, logistical, and human behavioral dimensions of wildland firefighting. Overlooking these forces was as potentially threatening to people's safety as any blowup. Thom developed his own LCES, which included a checklist of questions such as "Am I mindful of my well-being and in a good spot with focused clarity?" and "Do you have a Safety Zone in which you can communicate honestly?"

When I spoke to Thom, the twentieth anniversary of the Nuttall Fire was just months away. In those years, the incident had been whitewashed, he told me. Because no one had died, the shelter deployments were considered merely "precautionary." Everyone on H4 even received an award for their keen insight in utilizing the "tens and eighteens" and LCES. But one of the 10 Standard Firefighting Orders is *Know what your fire is doing at all times*. "We didn't know what the fucking fire was doing. It made that huge downhill run," he said. "They didn't really take ownership of the decisions that were made that day. Even me. I could have said, 'No, we're not doing this.' So, that's a bad environment. The near-term effects of that fire were, everyone was cool. The long-term effects are that it fucked a lot of people up, and they're still living the memories from that one shift. It wasn't necessary. Shit, it's been twenty years, and it still bugs me."

AFTER LISTENING TO Thom speak on the podcast, Mike locked the door of his office and sobbed. It was the first time he admitted to himself that he might have PTSD. The next year, he left the fireline. But he still couldn't imagine a different life. "I felt very out of place in the real world, like I couldn't have a real identity outside of fire even if I wanted to," he told me. He became a dispatcher at the Susanville Interagency Fire Center, mobilizing firefighters and resources to respond to wildfires and local 911 calls, including suicides. He worked long hours, and the stress of the job was enormous. Mike sought counseling through the Forest Service, but the therapist he saw had no experience diagnosing PTSD. Cassie, feeling desperate, found a therapist in Reno who specialized in treating first responders. Mike began driving 180 miles roundtrip to attend therapy sessions. His therapist encouraged him to consider a new career. "I couldn't really heal if I stayed in fire," he told me.

In August 2020, West finished an eighteen-hour shift at his job and sent his letter of resignation. After seventeen years, his base pay was $22.80 an hour. "In my career, I was almost burned over four times," he wrote. Still, "nothing has been more a threat to my life than the symptoms of PTSD." Mike focused his criticism on the lack of mental health education and resources for wildland firefighters. "Even though I have PTSD, I don't think I'm dangerous or crazy," he wrote. "I think wildland firefighting is dangerous and crazy and PTSD is a normal reaction from the human brain." A month later, he started work as a middle school teacher. But school was canceled during the first week, when the Sheep Fire threatened Susanville. The fires were still coming to him.

I asked Mike if there was ever a point when he questioned the effectiveness of wildfire suppression itself. "I had a lot of thoughts like that," he told me. "It would depend on the circumstances. Like, if I was in a fire close to any kind of community, it never crossed my mind that we shouldn't go put this thing out. [But] I was on a lot of fires way out in the wilderness where I thought, Well, if somebody gets hurt or killed out here, this is really dumb. This is kind of for nothing, this is a lightning fire. This area needs to burn."

After a while, Mike began to see suppression as a lost cause. "I almost felt like the generations before us had screwed up, they were suppressing fire so hard." Over the course of his career, he estimated that he had worked on hundreds of prescribed burns and never lost control of a single one. He saw firsthand how wildfires that were ripping through the forest crashed into those previous burns and lay down. But there was always a chance that a prescribed burn could escape and go huge. "It's all kind of hopeless, like damned if you do, damned if you don't."

"Do you feel like over the course of your fire career, climate change was becoming more of an issue when it comes to fire behavior?" I asked.

He replied:

I think the first time climate change really hit me and scared me was after the 2013 fire season ended. It was cold here, but we weren't getting any precipitation, and I got asked if I wanted to do an assignment in December around Christmastime. We threw together a crew and went down to Los Angeles to stage there. Then we got called to a fire in Big Sur on the coast, and we were driving up there at, like, ten o'clock at night. We come over the hill and this fire is crowning. I'm just going, "Oh my God. It's Christmas in Big Sur, it's supposed to be dumping rain, there's not supposed to be fires here." Then, in January, I went to a fire on the Lassen National Forest that went for fifteen hundred or two thousand acres in an area where there would normally be several feet of snow.

After years of intermittent or nonexistent snowfall in the winter, Mike started getting scared in the fall, wondering whether the snow would come. How dry could it get? What kind of fire would result? "I used to think I was a pretty good predictor of what the fire behavior is going to do, where's the safe place to be," he said.

And then I started being on fires where I'd be sitting in the safety zone when it was really ripping and thinking, Is this really a safe place to be, or are we going to get one of those big fire tornadoes? It has me questioning what is safe and what isn't. The fires in Napa in 2017, they just got up and went into the community, like, deep into the community, not just where the fuel is. It just kind of came into the concrete and was getting houses. What the hell? My house is in the middle of town. I don't know. When we had fires last summer, it had me thinking, are we going to get some kind of weird fire tornado? It's strange to think that, because if you asked me ten years ago, if my house in the city limits of Susanville was threatened by wildfire, I'd have told you absolutely 100 percent never. But last summer I was kind of thinking, I don't know. Maybe.

I said, "Thinking about PTSD and the fact you could be going out to work and coming home and feeling under threat from the same forces you are out there trying to mitigate is just crazy. Because then it feels like there is no safe place, and, to me, it seems like it would compound the anxiety."

"I've been with people on fires and they're getting a call from their spouse because there's a different fire threatening their home, and their spouse is asking, 'When do we leave? Should we pack up?'" said Mike. "So, here's this person three hundred miles away trying to put out this monster fire; meanwhile there's another one coming towards their home. I'll never fully break up with fire because it's here, it will always be a part of my life. I hate comparing anything to the military. But... it's like the front is here."

ON JULY 13, 2021, two weeks after Mike and I had that conversation, a power line went out near the Cresta Dam on the North Fork of the Feather River in Butte County, California. It was seven a.m., and the electric company PG&E issued a work ticket for a lineman in Chico to check on the outage. When the lineman arrived at the dam four hours later, he could see a blown fuse on a power line hundreds of feet up the canyon's treacherous slope. But he couldn't access it. Between him and it was the Feather River. The worker drove back down Highway 70 to the town of Pulga, crossed the river, and then drove eight miles up a one-lane dirt road that passed by the spot where, three years earlier, a PG&E power line had sparked the Camp Fire that killed eighty-five people. A couple of miles from the blown fuse, the lineman was halted by a road crew repairing a bridge and had to wait for hours to cross.

By the time the lineman finally reached the site, he found a tree lying over a live power line and a small fire—a "start"—measuring about forty by forty feet and slowly spreading down a slope. Using a fire extinguisher and hand tools, he tried to put the fire out alone. A handcrew was dispatched but wouldn't arrive for hours. An airtanker arrived and poured fire retardant spray, and a helicopter began dropping water. As the day turned into night, the fire continued to burn. Over the next twenty-four hours, it grew to nearly two thousand acres. This was the beginning of what would become the largest single wildfire in American history.

Cresta Dam is seventy-five miles away from Susanville. Mike didn't seem too worried about the conflagration that would be dubbed the Dixie Fire. I wasn't either. For the Dixie Fire to threaten Mike's hometown, it would have to burn through hundreds of thousands of acres across four counties and two national forests, jumping highways, rivers, and suppression crews the whole way. It would have to burn across the crest of the Sierra Nevada, something a wildfire had never done before.

Then it did all those things. In one twenty-four-hour period, the Dixie Fire grew by a hundred thousand acres. It sometimes sent up multiple pyroCbs in a day. Thousands of miles of dozer line, six thousand wildland firefighters, millions of dollars of fire retardant couldn't prevent the Dixie from marching toward gigafire status.

A month after the fire first ignited, Susanville was under an evacuation warning and Mike was packing his car under an ominous neon-orange sky in case his family needed to flee.

In early August, when the fire was a few hundred thousand acres in size, I started making plans to join a handcrew on the Dixie Fire. My conversations with Mike had convinced me that I would never understand the work of fire suppression until I tried it myself. Ben Wheeler, the wildlife biologist in Nebraska, gave me an idea of how to do it. Every summer, he went to California to do a two-week "roll" with the crew of a private wildland-firefighting company under contract with federal agencies. He did it to maintain his wildland-firefighting qualifications, but I sensed that the experience served another purpose as well. On prescribed burns in Nebraska, Ben liked to get close to the fire, watching and following its progression. I had often seen him standing quietly on a hilltop with his wrists folded over the hilt of his tool, perched like a sentinel as flames burned through the grass around him. His comfort with fire was striking and also confounding; many times I witnessed him emerge through heavy smoke like some sort of inflammable apparition from a direction where the fire had just burned. Going out west, I figured, was another way for Ben to get close to fire, another form of inquiry into its nature. I decided to follow in his footsteps.

6   DIRTY AUGUST

IT WAS THE HOTTEST SUMMER EVER RECORDED IN CALIFORNIA. THE ENTIRE SACRAmento Valley was under a heat advisory, and a blast of thick, hot air hit me as I walked through the automatic doors of the city's airport, a fire pack and duffel bag hanging off my shoulders. Dan Kelleher pulled up to the curb in a maroon sedan and jumped out wearing cargo shorts and his signature straw fedora. It was strange to see Dan out of his work truck and in civilian clothes. But, as he explained, he had no plans to get near the fire. We started the drive north to Chico, and he described defiantly ignoring what was happening and focusing on his daily life—training young wildland firefighters, attending Alcoholics Anonymous meetings, paddleboarding. Indeed, Dan was spending every moment he could on water, as though the Dixie Fire and he had opposite magnetic charges. Dan had grown up in the same mountains and rivers that the Dixie Fire was now clobbering, whispering to his parents before dawn as they lay in bed to ask for permission to be back by dinner. "I don't want to see it. I don't want to face it. It's getting close to where I spent my summers as a kid. I don't want to see that go." Yet one of his old haunts, the historic gold-mining town of Greenville, had already burned to the ground ten days earlier, leaving behind pulverized houses and the charred husks of cars.

"I just hit a wall, just feeling sad," Dan said, a cigarette resting between his fingers as he steered. "I took a look at the news, saw one picture, and said, 'That's enough.'" He paused and looked out the window. "That one hurt me." Even Dan, who I assumed had seen everything in his career, was taken aback by the severity of the Dixie Fire. "It's spotting like crazy everywhere," he said. Then he wondered aloud if he was too distant from what was happening in his homeland. "Maybe I'm turning away too much."

We drove toward Chico past mile after mile of fruit orchards and agricultural fields, an empty Pepsi bottle in the cup holder of the car for cigarette butts. Dan told me he had been on the Bootleg Fire in Oregon a couple of weeks before. The fire had ripped eastward—on its fourth day it grew by sixty-four thousand acres—and had sent up pyroCbs nearly every afternoon. As it headed toward the Sycan Marsh Preserve, a thirty-thousand-acre wetland with upland forest of ponderosa pine in the Klamath Basin and the heart of the Fremont-Winema National Forest, Dan had driven up to help out.

We pulled off the road and parked the car to stand in the shade of a peach orchard and smoke. Dan continued his story. One day, he was working in an engine and went to assist with a lowboy, a semi-trailer. As they were driving out behind the machine, he noticed that the wind was picking up. Through the windshield of the engine, he could see sticks, pine cones, and leaves flying through the air. But the debris wasn't going in the right direction. Instead of blowing away from the main fire, it was being sucked toward it. Dan was witnessing updraft: the fire was burning so hot and intensely that it was creating a pyrocumulonimbus whose convective energy was now sucking the warm air around it upward into the atmosphere. Dan knew the fire was big, close, and moving in their direction.

"For about thirty seconds, we were caught in the updraft of this running crown fire. I'm going to say it was forty miles per hour, but if someone told me it was seventy, I would believe it. I thought, Well, maybe this is it," he said. They managed to avoid a burnover by squeaking through before the front hit the road behind them. "It was an ass-biter," said Dan.

At times, the Bootleg Fire's updraft was so strong it uprooted mature ponderosa pines. But already reports were coming in that in places where forest restoration efforts had taken place—via prescribed fire and mechanical thinning—the Bootleg's intensity had been moderated. As the fire manager at the Sycan Marsh Preserve said, "It goes to show that if you reintroduce the process of fire, you start to chip away at that hundred years of fire suppression. You can create resilience in this forest type, in the sense it can stand up to a big, bad fire like the Bootleg Fire."

When we arrived in Chico it was 106 degrees. We drove to Zeke Lunder's house. In Zeke's living room, one of his sons watched a computer screen showing a live map tracking air tankers taking off from the municipal airport to make retardant drops on the fire. There were so many airplanes taking off and landing it looked like a game of Space Invaders. Zeke's eyes were glassy with exhaustion. He had barely slept the night before, staying up to analyze heat maps generated by the European Space Agency's Sentinel-2 satellite. They showed explosive growth of the Dixie Fire driven by massive thunderheads and wide flame fronts bearing down on highways. At one point that day, there were four pyroCbs in northern California, two of them generated by the Dixie Fire's front. At 2:30 a.m., when the fire finally seemed to be quieting down, Zeke had still been posting updates. "The Dragon takes a nap," he wrote. Zeke was analyzing satellite imagery, a process that involved assigning wavelengths of light different colors in order to evaluate the burn severity of the fire, but his initial analysis was dumbfounding.

"I am not used to seeing such a high proportion of the landscape burned with such high severity," he wrote.

In early August, Zeke had begun sharing intel he was gathering from the Forest Service's fire maps, infrared (IR) maps, and satellite imagery on a blog he created called *The Lookout*. Every day, he provided analysis of the fire's behavior, historical context of previous wildfire burn footprints and logging operations, and predictions for what might come next. *The Lookout* gained an instantaneous readership of local community members hungry for the type of granular information on the fire's progress that only someone like Zeke—a geographer, an expert in geographic information systems, and a native son to the area—could provide. The blog was a source of information but also a place for unfiltered commiseration. Zeke's brother and parents had both been evacuated. "Well, things are really going to hell on the Dixie Fire," Zeke wrote one day. On the evening that Greenville burned, he wrote, "I don't really know how to write about any of this. Today was terrible for everyone involved—from expats living away watching the awful IR webcams to the people who had to run for their lives. The losses in Plumas County are pretty staggering and personal." Zeke considered the loss of Greenville to be a historic and generational event for northern Californians that would be remembered in the same way as the Big Blowup of 1910, the Loma Prieta Earthquake of 1989, and the 2018 Camp Fire.

As we talked over dinner, the consensus was that the Dixie Fire was doing things that few people had ever dreamed a fire could do. For decades, the basic tactic of wildfire suppression had been backfiring—lighting fires along ridges, roads, and bulldozer lines ahead of the main fire. These operations typically took place at night, when the fire lay down, because that was when temperatures cooled, winds died down, and humidity rose. But, as Zeke explained, nighttime on the Dixie Fire was different. The fire didn't lie down; it just kept running, flanks morphing into heading fires that spread east and west. Fuels were so dry and winds so strong that the fire was charging through drainages and sending up embers a mile ahead that sparked spot fires that turned into new conflagrations. There were so many spots that perimeter maps of the Dixie Fire looked like splatter paintings. Even three-hundred-foot canals that would normally act as fuel breaks weren't stopping its progression. Hotshot crews often couldn't go direct and attack the fire with handlines. Even bulldozers were struggling to get close to put in line. Sometimes the firing operations were spotting and escaping, effectively enlarging the size of the fire rather than containing it.

Zeke explained that hotshot crews typically revered for their burning skills were now struggling with no good options. Firing operations that had contained parts of the western edge of the fire had also ignited a crown fire that burned ten thousand acres of old-growth forest, including a rare creek where Chinook salmon still made annual runs. "It's easy to put too much fire on the ground, and once you light it, you can't take it back," he wrote on *The Lookout*. It was so bad that some people were wondering if firefighting tactics were making the situation worse. At a recent work meeting, Dan told me someone had asked, "How big would this fire be if we *hadn't* fought it?"

That night, I slept on the outskirts of Chico, filled my pack's water bladder and two metal canisters—over ten pounds of water—and laid out my pants and boots. I had my own fire pack by then and had spent the previous weeks wearing it on sweltering hikes in the desert around Tucson, Arizona, with my four-year-old son, who happened to weigh exactly forty-five pounds, straddled on its top. Before I left, Bre Orcasitas, a former hotshot, rappeler, and smokejumper, had sat down with me in her living room in Tucson and spread out my gear on the floor. She inspected each Ziploc bag, bottle, and item of clothing for practical value and weight. She lent me a pair of Nomex pants that she had altered herself for dangling upside down out of helicopters and that she deemed more durable than my own. She gave me a spork and a space blanket. I couldn't understand why I needed the latter. Wasn't it going to be hot near the fire? A week later, when I found myself shivering in my sleeping bag and desperately wrapped the space blanket around myself for extra insulation, I sent my thanks to Bre up into the night sky.

I ARRIVED AT a gravel parking lot near the Chico airport in the dark. In my pack was a "red card"—an identification card with my incident qualifications and photo on it—that had been issued to me. I could barely make out faces in the headlights as the crew threw duffel bags, fire packs, tools, chainsaws, and pallets of water and Gatorade into the beds of white Ford Super Duty trucks. I shook hands with the crew boss, Gene Lopez, who wore a sweatshirt and a ball cap emblazoned with the private contracting company's logo pulled low over his brown eyes. The crew was a Type 2 Initial Attack crew. In addition to constructing handline, its members included multiple qualified incident commanders, or "squad bosses," who could lead modules of four to six people on initial attacks on fires. In order to maintain IA status, a crew has to have three qualified sawyers, and at least 60 percent of its members have to have more than one season of experience.

Gene and Dan knew each other, and now they stood around catching up. In July, the crew boss told Dan, the crew had been on the Beckwourth Complex, two lightning-ignited fires in the Plumas National Forest. The fire's effects had alarmed him. At crew briefing one morning, he picked up a handful of soil and ash at his feet and showed it to the team as a warning. They were at a high enough elevation that moisture should have limited the damage caused by the fire. Instead, everything around them had been reduced to powder. "Look at this. It's nuked at seventy-five hundred feet. That's what we're dealing with," he told them. "I'm like, 'Dude. We've never seen this before. Everything is aligned and receptive to what is right in front of that fire.'"

Dan shook his head. "They're putting three to four times more logs on the trucks in Oregon because the wood is so dry," he said.

Gene told Dan that during their previous roll, the crew had raced a downhill crown fire into the town of Greenville. Then they had to leave the town itself when it was hit by fire and destroyed.

"It was the first time I had to say to my crew, 'Head out, let's go,'" he said. "I'm responsible for this crew. I saw a downhill run over the ridge that I would normally see only running uphill."

"It's good to check your internals," said Dan pointing to his head.

I couldn't smell smoke, but as the sky began to lighten around us, I could see the blue-tinged fog of particles from the fire, which was fifty miles away. I was to be the fifth member of the crew's "Charlie" module, and I threw my duffel and pack in the truck bed and sat in the back of the cab between two wildland firefighters in their twenties. We pulled out of the lot in a caravan, the four trucks in tight formation as they picked up speed on the highway.

I was prepared for skepticism, even hostility, joining a handcrew in "dirty August," the halfway point of the fire season, when wildfires are typically ripping and crews are working so hard that days blur together. People are sleep deprived, overworked, and really, really dirty. As the host of the podcast the *Hotshot Wake Up* described it, dirty August is also when the asshole in people comes out. "Perpetual dark sarcasm is like constantly describing your world as inverted," he said on his podcast. "To constantly be surrounded by an inversion of the truth weighs on people. That's why August is dirty."

By the time I showed up, the crew was already on its ninetieth day of fighting fire. No one was unfriendly so much as they were completely indifferent, too busy and tired to care much about why I was there. People were way more interested in sleeping. In the truck, everyone fell into a deep, heads-lolling, narcotic-like stupor. The head of the twenty-six-year-old firefighter to my right, a guy named Garrett Rangel, kept falling on my shoulder until he would abruptly snap awake and apologize before doing it all over again.

I was too apprehensive to sleep. I tracked our route on a map and looked out the window, where I could see a fire scar, denuded trees sticking out of the hillsides like medieval spires. "What fire was this?" I asked the driver, a guy named Jayson Ramsey who was also the crew's captain. He had a salt-and-pepper beard and blue eyes.

"Camp Fire," he said.

As we turned north onto Highway 70, the early sun was a hot-pink orb. We stopped at a gas station, and people picked up microwave burritos, coffee, dip, snuff, cigarettes, sunglasses. Back on the road, we passed the North Fork of the Feather River, where the Dixie Fire had originated a month earlier. We were now, essentially, following the pathway of the Dixie Fire, chasing the dragon's tail. As we made our way up through the rocky hills and steep valleys of Feather Canyon, we began passing signs that read, "Thank you heroes" hanging on now vacant houses. Just after we entered Plumas National Forest, the cell reception cut out, and we drove through tunnels blasted through the rock alongside the Feather River. The road twisted and turned, hugging the side of the steep canyon wall. Jayson started pointing out handlines dug by crews on the slopes that were now scorched by the fire had that blown past them. A single thought kept appearing in my mind: How did we fuck things up this bad?

"Wonder where they're going to put us," said Jayson.

"Bet we're going to Susanville. We haven't been there yet," said a guy named Frank sitting next to him. Frank had a missing front tooth, thanks to an unfortunate encounter with a tree last season, he told me.

"So," I asked, trying to make conversation, "did you guys all get trained by Dan?"

Jayson looked over at Frank and they smirked. "Uh... no," said Jayson. "Some of us got trained in other places."

It took me a minute to realize he was referring to California's Conservation Camps, which are run by the state's correctional department. Each year hundreds of incarcerated people are trained and become part of inmate crews fighting fires. Maybe a conversation for another time, I figured.

I hoped that Frank was right about our going to Susanville. Mike West had told me he lived close to the fire camp there. Before I left, he had texted, "Just be careful out there. This fire is insane. I have a hard time processing how big it is and what it's doing. It's all over the map. It's like five giant fires."

Instead, we were assigned to fire camp in Quincy, a small town in Plumas National Forest. The town's fairgrounds had been transformed into an incident command post for the east division of the Dixie Fire. Hundreds of tents and trailers were scattered around, and the fairground's stables were now full of evacuated horses, goats, and chickens. The crew boss checked in, and the trucks got an inspection. Volunteers came around handing out plastic clamshells of expired blueberries. I watched sprinklers dousing thirty-foot-high piles of logs with water at the Sierra Pacific mill across the street. We talked about the recent arrest of a criminal justice professor who was charged with arson for setting fires near the Dixie Fire. Jayson watched a local news story about wildland firefighters on his phone. "You just watchin' a big documentary on how cool we are, bro?" joked Garrett.

We left Quincy and set out north toward the Indian Valley. Our assignment for the day was to do point protection on homes, prepping for the coming fire by digging perimeters of bare dirt around them. On the way, we passed through places the fire had already burned through. Some homes were incinerated, while, next door, another home still had baskets of flowers hanging from its porch. "It picks and chooses," someone said.

We stopped at a drop point in the woods to put on our yellows and gathered up PPE and tools. "Fire's being checked up right now in Peters Creek, so we've got a couple days to get it ready," Gene told us. "They lost houses last night and the night before. Then fire did a 180 on 'em, came back at 'em, and took the rest. It's all the same thing that we were facing when we were here last. Pay attention."

The entire valley sat under a thick smoke inversion, in which a cap of warm air trapped cooler air and smoke low to the ground; the mountains around us were invisible in the pall. The temperature was a hundred degrees, and the Air Quality Index was 368—a "hazardous" rating. The milkweed on the side of the road was drenched in psychedelic-pink fire retardant. The first house we went to had been evacuated. We removed piles of firewood stacked alongside a house whose deck was lined with incredibly detailed wood carvings of eagles and bears. On a table on the deck, an opened but undrunk bottle of Budweiser sat next to a drill, a staple gun, and a balled-up American flag. At the next house, we pulled away bags of trash, old lumber, scrap metal, and children's bikes from the sides of the house and then began to dig line around it. Someone had given me a Pulaski from the truck, and as I swung the heavy, clunky tool it sent up puffs of dust from soil so dry it was like sawdust. I began to sweat. My heart beat faster. The heat felt so oppressive it was like being suffocated. My breath became rapid, and then, all of a sudden, it felt like there wasn't any oxygen. A wave of pure panic flooded my brain. I was sure I was about to die. I couldn't believe it. It was the first day on my first roll digging my first line and I had failed. In that moment I had the very lucid epiphany: I was a fucking idiot for going to California.

Somehow, through strength of will or fear of humiliation, I continued to swing and dig, moving down the line, fighting my panic and the urge to break line, until I reached the end. When I finished, I stepped away from the crew and waited for the adrenaline to fade. My breathing slowed. Back at the trucks, one of the squad bosses saw my Pulaski and exchanged it for a rhino. "This is better for you," he said. The tool was like a shovel with a blade at a right angle for cutting and scraping. It had a longer handle and it could move more dirt with each swing than the Pulaski. At the next structure, a giant barn, I worked faster and didn't lose my breath.

By the afternoon, the cows were like shadows in the thick haze hanging over the flat bottom of the valley. There was no horizon. It was like a camera exposure had been left on and now the world was blurred at its edges. We had been under the claustrophobic inversion for nine hours. "I prefer choking on smoke than this," said Jayson, looking out the truck's window. "This messes with my head. It's like being boxed in."

In the waning hours of the evening, the crew mostly waited inside the idling trucks, where they could breathe conditioned air. I sat in the bed of the truck against a pile of duffel bags. I tried to understand the panic I had experienced earlier. The setting had been pretty benign. The nearest fire activity to us was to the northeast, where it had made a run the night before. Was I fire-scared, or had I just spent too many hours over the previous months talking and thinking of nothing but extreme fire behavior, spending hours watching entrapment videos on YouTube, poring over incident reports on some of the highest-fatality fires of the last two decades? Maybe I just wasn't acclimated to the elevation. But I worried that if I couldn't handle digging line around houses, I would be incapable of digging line around actual fire. I listened to turkeys calling somewhere out in the smog and talked with people as they grabbed snacks and water and Gatorades or stood around smoking. I decided to stay.

The crew had a misfit vibe. Frank was in his early fifties, which was pretty old for a handcrew. He had black hair that for years he grew down to his waist but was now shorn. The young guy sitting on my left in the truck was quiet, but I eventually learned he'd run away from home as a youth and lived under a bridge for a while. He now had a young son. His parole officer had told him about wildland firefighting. He liked to cook his ramen noodles on the tailgate and add mayonnaise, bottled lime juice, and crushed hot Cheetos, a recipe he developed from prison commissary ingredients. Garrett was the youngest. He had been raised in a fundamentalist Christian family near Tahoe and was now a self-described nihilist. There was one other woman on the crew—it was her rookie year—but she was in another module, and I barely saw her except for in the evenings, when we ran into one another in the empty women's shower trailer at fire camp.

I learned one of the crew's former sawyers was serving a life sentence in prison. The previous crew boss had left to fight against Azerbaijan in the 2020 Nagorno-Karabakh War. My squad boss had hoped to replace him, but a catastrophic, drunken motorcycle accident in the off-season had required months of rehabilitation and ended his prospects. A lot of the crew members had learned their trade in California's camps. Prison fire crews sounded miserable. People would laugh hysterically remembering the shitty and dangerous assignments, or how they ran the saw for so many hours their bodies would curl on the ground from muscle spasms. But I couldn't detect much resentment. "Camp saved my life," one crew member told me. He was a military veteran who had gone to Eel River Camp. "I would have never gotten here if it wasn't for it."

Actually, I never talked to anyone on the crew who didn't love wildland firefighting.

"I don't ever want to do anything else," Garrett told me as he ate an MRE from a box in the truck bed.

"Fuck, I was working in a factory before this," added someone. "I just don't want to be inside."

IN THE MORNINGS, my tent was often covered in flakes of ash. I put on my boots, stiff from the cold, peed in a porta potty, and brushed my teeth at a row of outdoor sinks. Then I set out to walk through fire camp before morning briefing started at seven a.m. Camp was like a pop-up, military-style, disaster-response city. Our sleeping area was an old racetrack that was also designated a safety zone should the fire blow through town. I walked past a supply warehouse and a medical trailer and toward the central camp, where dozens of trailers for showers, laundry, food, and the incident command were located. My route took me past an old clapboard building sponsored by the Plumas National Forest. A wooden sign hung off its upper deck, and its message never failed to strike me with its irony: "The strength of any nation depends upon its conservation and wise use of natural resources."

At morning briefing, I stood near my crew and squad bosses listening to the incident commanders deliver their updates on the fire's progress from the stage, watching the hundreds of people around me shuffle and drink coffee. Everyone seemed to wear the same uniform: a crew sweatshirt emblazoned with a logo or the name of a big fire they'd worked on. You could often tell the hotshots from their pants, which were cut high at the ankle. I flipped through the daily Incident Action Plan, which listed air operations, fire-behavior forecasts, and safety messages. It was sixty-six pages long. Probability of ignition: 100 percent.

Early on, our assignment was to mop up, also known as coldtrailing, on the northern edge of the Indian Valley. The fire had burned through there a couple of days before, but the ground was still holding heat. Ash pits were created by smoldering embers that ate away at the underground root systems. Our job was to grid—stretch out in a staggered line—and make sure at least one hundred feet of ground from the perimeter of the fire, often a strip of bulldozed earth, was completely cold: no smoke, embers, or flame. Unless we called in an engine for help, we had no water and relied on chainsaws and hand tools. It might take three or four mop-up patrols to ensure a place was cold before the status of a fire perimeter could change from "active" to "contained."

We drove the trucks up twisting, unnamed dirt roads into the heart of the woods. The crew split into two and leapfrogged one another as we hiked ridges alongside a warren of tracks from bulldozers that seemed to have crashed through like giants, ripping up roots and tipping over mature trees as though they were blades of grass. The soil in the tracks was orange and as soft as talcum powder. We climbed up and down these exposed innards of the forest that were so steep at times it seemed like the dozers had defied the laws of physics. Midslope, I sometimes stopped climbing, my feet awkwardly splayed in an attempt to get a purchase on the loose dirt, and marveled at the person who possessed the courage to drive a twenty-thousand-pound metal machine along the edge of an active fire at such inconceivable angles. Then, as I began to crawl up again, my feet slipping a foot for every two I hiked, I cursed their bravery.

Mop-up, Garrett told me, was his least favorite part of the job. He considered it the most dangerous because you had to be constantly aware, careful not to step into an ash pit on the ground or get hit by fire-weakened trees called snags falling from above. We worked spaced out from one another, scanning the ground as we walked. It wasn't always obvious where the heat was located. At first, I checked by bending down and using the back of my hand to feel the dirt. Frank saw what I was doing and showed me how to use my tool to scoop up the earth and bring it to my hand instead. It was OK to bend over, but after ten hours I would regret it. Sometimes I would scoop up what I was sure was cold, brown earth and be surprised to feel that it still held warmth. Sometimes I saw places where fire had burned so hot it turned an entire tree to "moon dust"—white ashes—that I assumed was warm only to find it cold.

While gridding, any message or instruction was sent up the line from person to person like an echo until it reached the furthest person—often Frank, who seemed to prefer working deep in the woods—who would then belt out the message at the top of his lungs like a carnival barker, confirming that he'd received it.

"Watch-out-for-widow-makers-and-stumps-last-man-got-it!"

"Moving-last-man-got-it!"

"Hold-for-heat-last-man-got-it!"

Getting the heat out of the ground mainly involved exposing the soil to the air. We opened up the pits and reached into the deep tunnels to rake out embers with our tools. Then we chopped the embers into tiny pieces or smothered them until they died, and then watched for any wisps of smoke. Sometimes Frank would find a hot spot and grab a plastic water bottle from his pack, take a gulp, and then slowly spray it over the spot to cool it. "Don't you ever do this," he warned me. We scraped away the burning char on logs or dug cup trenches around them so they couldn't roll and catch nearby detritus on fire. It was filthy work that turned our hands and faces black. It helped if you were a little bit obsessive, chopping every last ember into little pieces, creating perfect hillocks of dirt for trenches.

One afternoon we came upon a burning stump deep in the woods. Nearby, a stream burbled. Someone suggested using our hardhats to scoop up the water. We formed a line, ten people passing their hats up and down in an improvised bucket brigade.

"This is some hero shit," muttered one.

"You know how many kittens we just saved?" added another.

There were chatter and insults and laughter, but for long stretches of the day we were spread far enough apart that it was quiet. I could almost imagine I was there to take a pleasant walk through the woods, which managed to still be beautiful. Then an emergency evacuation alert from a nearby town would hit all of our phones at once, a screeching alarm pinging around the forest, reminding us that fire was still blowing up. The alarms motivated me to attack the embers with more energy: if winds shifted or intensified, they might reignite.

During mop-up, nature's comforts became unsettling and sinister: a gust of wind cooled the sweat on my face, but I worried that it might awaken embers; a beam of sunlight pierced the smoke, but I wondered if it might portend a lifting of the inversion—a change in the stability of the atmosphere that could bring unpredictable fire behavior. I spent a lot of time on breaks scouring the incident maps and the red perimeter of the fire. Where was it? When would we see it? Did I want to? We were close but somehow always circling its flaming tail as it slipped around a corner or over a neighboring ridge or snuck behind us. "Those houses that we dug around yesterday?" the crew boss said. "They're getting hit right now."

We moved through the forest at a pace that seemed slow to me. I had yet to learn the trick of conserving one's energy for the moments that counted: attacking a spot, digging hot line, running, nightshift. At every opportunity, even during a ten-minute break for water, people slept. They napped on the ground with packs or even rocks for pillows. I watched in wonder as a sawyer napped while he stood leaning against a tree. One day, I was hiking across a sun-drenched ridgetop when my mind suddenly came to, as if I had been startled awake. Was it possible to doze while walking? Hiking up and down through one hollow and ridge after another for hours and then days made the work feel dreamlike in its monotony.

"I hate to tell you this: it's either like this or you're *in* it," someone said to me.

One day Coby Howarth, a crew veteran, showed up. Coby's wife was from Kenya, and he had been trapped there during the pandemic. The crew was ecstatic at his return. He had red hair, freckles, and glasses, and he was massive. He blamed his weight on eating *ugali*; his yellow seemed to be made with enough Nomex to construct a tent. Coby's role on the crew was that of philosopher-jester. He didn't care about playing it cool or tough. He would hold everyone rapt with an earnest rendition of the Bee Gees' "More than a Woman" in the middle of the woods, opine on the ways the film *Inglourious Basterds* illustrated the principles of ethnomethodology, or keep everyone bent over in laughter with bloviating messages on safety at morning crew briefings. "The trees are taller than you. That means they're dangerous. If the fire is taller than you, that means it's also dangerous," he would drawl. "If [fire's] on the ground, it might be OK. Might be able to do something. My advice to you, maybe if you're running with the saw today, if he cuts anything, maybe fill it up for him. He might be a little tuckered out. May be a little tired. I think the wind is shifting every two or three hours? That's what I got from the weather. Could be good. Could be bad. I don't know. The best solution is going up there and seeing what's going on."

Coby had come to wildland firefighting by way of a crackpot idea hatched during a spontaneous hike with friends through Death Valley. Now he used the money he earned each fire season to buy the cheapest possible plane tickets to visit his wife, booking itineraries that involved multiday slogs with layovers in obscure cities. This summer, he was back with the crew to lose weight, make money, and "smash."

"Smashing" was one of Coby's favorite words. It basically meant working one's ass off in the woods. But it captured some vital aspect of wildland firefighting that Coby seemed to believe was at the heart of what they were all doing out there, which he explained to me in detail over the days as we rode in the back of pickup trucks or stood around an ash pit taking turns hacking at the embers in the ground.

"Essentially, you have to have integrity. That's basically what you gotta have out here," Coby said. "And you basically just got to have an indomitable spirit. Like, you have to be able to just keep going even if, like, you are really uncomfortable or you're in a bad situation."

"Do you have to like uncomfortable situations?" I asked.

"You might even want to seek them out," he said. "You might even like the prospect of doing something that's incredibly uncomfortable. That might even be appealing to you. Like, you guys want to go to some dangerous country and hang out? Awesome. These trees are on fire. Better cut them down."

The culture of wildland firefighting, according to Coby, was rooted in a few core practices into which everyone was indoctrinated. First, all wildland firefighters swear their allegiance to the *Incident Response Pocket Guide*, the small, spiral-bound notebook issued each year by the National Wildfire Coordinating Group that contains all the critical information for operations—from watchout situations to how to select a helicopter landing area. Second, all wildland firefighters learn and speak a shared language. "If you noticed, on radios it's very difficult to communicate anything," explained Coby. "You have to be kind of laconic, you have to try and use as few words as possible. But you have to actually say something.

You can't be like, 'Hey, do the thing. Do what has to be done.' You have to be skills-specific. You have to change your language."

Third, wildland firefighters thrive on grueling physical labor. Smashing. "I feel like it satisfies something in my brain and my psychology," said Coby.

"I for sure need it," added Garrett. "I'm a mess without firefighting. I have an energy, I don't know what to call it."

"I'll just have, like, an abundance of energy that I have literally nothing to do with," said Coby. He turned to me. "Wait till you get home. You're going to be sitting on your couch and your just gonna be, like—it's not jittery or that you can't sit still. You'll just feel, like, this white energy. You'll feel full." That feeling was how you knew it was time to smash.

Since Coby had taken my seat in the back of the truck's cab, I now sat in the front row between Jayson and Frank. My job became catching the radios, snacks, cigarettes, phones, and containers of dip as they rolled off the dashboard at each turn, and constantly refreshing Jayson's phone in places with spotty cell service so he could watch the documentaries and horror movies that he streamed constantly. There was zero privacy in the truck. Everyone knew when you needed to pee, what you ate, what mood you were in, the bills you had to pay. Phones would ring with kids asking to borrow money, buddies who had brawled the night before, updates from lawyers. One afternoon, I talked to Jayson's health insurance company, reading his member number from his card to the representative and relaying messages as he drove.

Jayson was intimidating. He had grown up deep in the California sticks (another crew member described it as looking like somewhere in Afghanistan) and had what he called a "piney" accent. He did not like to chat. I teased him that he had a "resting murder face." But he also had sparkly eyes and two daughters back home that he talked to on the phone and obviously adored. He had a childish love of picking up critters in the woods, baby rattlesnakes and lizards. One afternoon when our crew's patrol led us onto a paved road in the Indian Valley, we began walking in formation. A herd of horses stood and watched us and then broke into a canter, wind whipping through their manes as they ran alongside us. Jayson broke line and walked across the road to get closer to them, whooping and grinning.

Garrett told me early on that if things got hairy, I should follow Jayson's lead. He had seen a lot, and I would know if things were serious by how he reacted. "And," said Garrett, "if things start getting fucked, just stick with him. That's what I do."

"IT'S LIKE WE'RE in a lull," Gene said one morning. "It's a weird feeling on this one, waiting for it to blow up again." Late that afternoon, we got word that the fire had reignited and slopped over a perimeter near a place called Moonlight Ridge. The crew split, one half hiking up to the slopover from the valley and the other hiking down to it from the ridge. Whoever reached it first would do the initial attack. We parked on a dirt road and put our packs on. Garret told me to drink water until I had to pee and didn't want to drink water anymore, and then to drink more. "If that spot heats up, you'd just sweat all of the water out hiking out of here." Those of us in the hiking-down group got in line and started descending an incredibly steep bulldozer track; I dug my heels into the dirt to avoid falling. I could feel the wind blowing in my face. Huh, I thought to myself, an upslope wind. I strained to hear or see or smell the fire ahead. We were hiking into the hole, but I tried to locate fear in myself and couldn't find any. I wanted to find the fire.

After about thirty minutes of hiking downhill, we got word that the other half of the crew had reached the slopover first. We turned

Dirty August

around and made our way back up the treacherously steep track to the trucks, our feet slipping, sweat dripping from our faces.

"You OK, Coby?" someone asked at the top.

"I was feeling emotional at the bottom," he gasped. "But I'm good now."

Then instructions came over the radio for us to head back down into the hole. There was still so much heat on the mountain showing up in infrared maps that it needed to be mopped up. The crew groaned and hoisted on packs and went back down the now familiar bulldozer track. We spread out through the woods to dig root stumps and embers. By the time we finished, it was sunset. As the crew loaded up, I sat on the edge of the road and hung my legs over its steep ledge, looking over treetops tinged by fire as the hot-pink star slipped below the mountains. A root pit smoked below me, and more smoke blew past from a north wind. My phone shrieked: the outskirts of Susanville were under evacuation orders.

OTHER THAN MORNING briefings, I rarely saw Gene during my first week with the crew. I decided to take it as a compliment: he wasn't babysitting me. I would either sink or swim, and he would hear about it one way or another. One morning, we finished doing mop-up around a couple-acre slopover in a grove of towering trees, and the crew headed to patrol a dozer line through the forest. I walked over to the pickup where Gene was standing and asked if I could hang back and talk. "Sure," he said.

Gene sat in the passenger seat, and I leaned against the door. I had noticed in his interactions with the crew that he wore his authority lightly. He didn't badger or shout or yell often. He somehow communicated a lot by not saying much. It reminded me of good parenting; I knew he had four kids at home. Now he explained that his style of leadership was like the Wizard of Oz: he was always behind the curtain, watching, orchestrating, nudging, pushing when needed. He didn't like to micromanage. But, he told me, he made sure the crew members knew he had their backs. Rather than coercion, his approach used people's pride and expectations for themselves for motivation.

Gene had been trained in a prison camp, but he'd managed to make a career out of firefighting. I asked him whether this shared background with some of the crew helped him build trust with them. "I don't ever want these guys to think I act like I'm better than them," he said. "I'm here, I'm part of this crew. It's deeper than just work." He continued, "A lot of it does come under background. Maybe they had a rougher life at home. They were exposed to a lot of stuff at home. It made their shell very hard. When we go in and finish a line, hold a fire up, the reward they get for that is something they were striving for in their life. The relationship they might have had with their parents—they were never given accolades." He paused. "I tell them that I'm hard, they'll tell me I'm soft. I give these guys what they want, what they deserve."

Nearly every day on the Dixie Fire, we had driven back and forth through Greenville on our way to assignments. It was a haunting sight. The town was a graveyard of twisted metal and charred car frames. Exposed chimneys stood amidst the rubble like tombstones. Every time, I thought about Mike West, whose office when he listened to the podcast with Thom Taylor had been in Greenville. Garrett told me that the day the town burned had begun like any other: totally normal. They were watching the backdoor—an area that has just been burned—for a hotshot crew, making sure their fire didn't spread into the green. Then the crew boss had appeared in the forest. That was what set Garrett's alarm bells ringing. Most days, the boss was doing his Wizard of Oz thing—operating out of sight of the crew, monitoring radio traffic from the trucks, communicating with the division supervisor, watching weather reports, scouring topo maps. "We never see him on the line," said Garrett. "I'd never seen him yell. That day he was yelling, 'Go go go!'" Garrett showed me a video on his phone he had taken as they escaped the front of flames tearing through the forest. The crew is sitting in the bed of the truck as it speeds down a dirt track so fast the trees are blurred behind them. They were laughing and shouting, "We're going to die! We're going to die!" Fleeing for one's life, I figured, led to complicated feelings, even mirth. Also, fear. "I'm not going to lie," Garrett said. "I was scared. My heart was pumping."

Now I asked Gene, "What happened in Greenville?" The first thing he realized that day was that there was nowhere to park all the crew trucks. He and a couple of the guys had to drop off the crew in the forest and then shuttle the four trucks back down the mountain. They drove a single truck back, and the boss decided to hike the fireline to help keep an eye on things. The whole crew was stretched out over a mile, dropping individual trees that had caught fire, careful not to let them fall onto the green side of the line. As Gene hiked, he noticed a sixty-foot tree on the edge of the line whose top was starting to burn. Limbs were dropping onto the ground. The wind was increasing, and he worried that the burning debris would blow across the line and ignite the green. This tree's got to go, he thought. He called his trainee on the radio and said he needed wedges—pieces of hard plastic that help sawyers fell trees in the direction they want them to lie. But the wedges were back with the other trucks. Before the trainee left to get them, the crew boss told him to drive a second truck back with him. "I don't know, maybe it was my gut," the crew boss told me. "I said, 'Take a driver, grab a second truck.' It's easier to get twenty guys out in two trucks than one."

As Gene waited near the burning tree, he started noticing that the wind was now making its crown swirl in a circular motion. The fire was picking up, he thought. Then the division supervisor was on the radio telling him to round up his guys to leave. "Get down to pavement," he was told. Gene told everyone over the radio to return to the trucks. He counted the people he could see near him. Eight. There were still eleven people out on the line. He called over the radio to the guy furthest away: what was his head count of the people near him? Seven. There were still four crew unaccounted for. "You got to double-time this. We got to go, man, we got to get out of here," he said over the radio.

Soon after, Gene started to see the flame front coming through the trees. He told the people near him to get to the trucks, and he set out alone to find the rest. When he finally found them, he shouted for them to turn around. It was too late to make it back to the trucks; they had to get down the mountain on foot instead. They began running. "The way this thing's moving," Gene started to wonder, "how far can we make it before we have to actually think about a deployment?" Just then, he noticed a blur of green in the distance, moving through the trees. It was the hotshot crew's buggies racing out of the forest. He radioed to his trainee.

"Are you in front of the buggies or behind them?"

"Behind them."

"OK, slow down. When I yell, 'Stop,' stop. I think we're going to tie in with you."

Gene and the stragglers cut through a drainage and caught a glimpse of their white trucks through the trees. The boss yelled into his radio, "STOP!" They jumped in the back of the trucks and sped out. "When we hit pavement, we pulled across the railroad tracks and into a dirt lot," the crew boss said. "By that time, the fire was right there." Other wildland firefighters there that day reported fifty- to hundred-foot flame lengths on a fire that was backing down a slope.

As the crew drove through Greenville, they paused to put out a spot fire, but then there were so many spots that they were running around like maniacs. Gene told them to stop. "How do we know we're not going to get ourselves enclosed?" he warned them. They staged at the high school; when a local gas station caught fire, they were ordered out. The town was surrounded by fire. It was just a matter of time. "We call it a stand-alone," said Gene. "If it survives, it will only be by an act of God, because there's nobody there."

7 EVERYTHING IMAGINABLE HAPPENED

WHEN PEOPLE ASKED MY CREW BOSS WHAT HE DID FOR A LIVING, HE TOLD THEM HE was a dirtbag. Wildland firefighters fought fire by working in the dirt. It was invisible work. "We're not the ones in the big red engines. We don't do calendars," Gene said. "We're never seen."

But the nature of the job was changing.

"My first two years, I never saw a house, never saw an actual paved road," said the crew boss trainee. "The last three years, I spent more time in town than I did in the forest."

After the Camp Fire burned the town of Paradise and killed eighty-five people, the crew boss trainee told me he had been a part of the search and rescue, looking for human remains in the wreckage. "Having a twenty-man crew that has kids who are eighteen," he said. "[They've] never seen a dead body in their lives, and [I'm] telling them, 'Hey, we're gonna go through these buildings that are demolished, and you're going to see these cute, adorable dogs that go through it, and every time they bark that means there's a body there and we have to go in and find particles the size of your pinky nail.'" By the end of that roll, the crew was reduced to nine people. Everyone else had quit.

Gene's first close call on a wildfire was in Idaho. He told me it was one of his formative experiences as a wildland firefighter. In 2016, the Pioneer Fire was burning in Boise National Forest, and he was a sawyer for a handcrew, working on a ridge. Below them, a crew was conducting burn operations, lighting fires along a perimeter to contain the main fire. He watched as a column began to develop and realized that the crew below had lost control of their burn. The wind was now starting to push the column of smoke toward them. When the fire jumped a river, his crew began to make an escape, running for twenty minutes along a dozer line to get to a road. Just before they reached it, however, the smoke column lay down in front of them, and flames cut them off. Everyone pulled out their shelters, preparing to deploy if they needed to. All of a sudden, a man emerged from where fire had just burned over the road. "It was like a ghost, this guy comes walking out of the smoke," Gene recalled. "Everyone is like, 'Is that a *guy*? Where did you *come from*?'"

It was a wildland firefighter who had been line-scouting the fire, hiking the terrain to figure out where to try to dig line. He had been able to outrun the front for a bit, but when it finally caught up with him, he had found a drainage and bunkered in it as the fire blew past.

As the line scout told the crew boss this story, a supervisor came up and instructed the crew to head downslope and start cutting handline. This was alarming. "It went against everything we knew," explained Gene. "Fire down below, unburned fuel between us."

The line scout looked him in the eye. "From one friend to another, turn that down."

"I just met you," he protested.

"I'm telling you," the line scout said. "*Turn that down.*"

Everything Imaginable Happened

Based on the scout's advice, the crew refused to cut the line without a guarantee of air support. Gene never forgot how the sequence of seemingly random events unfolded and then coalesced into something like serendipity that might have saved his life. The same thing had happened the day they outran the crown fire outside Greenville. "A tree," Gene marveled. "Basically everything aligned and started things in motion to get us to where we're here right now because of a tree." He looked at his trainee. "I was gonna take that tree. I needed the wedges. You went down there, and as you're going, I said, all of a sudden, 'Hey, get a second truck.'"

We were still standing around talking as the light in the sky dampened and the crew emerged from the woods, slick with sweat. Then a young man on a dirt bike came tearing down a dozer track and pulled up next to us. He introduced himself and told us that a lot of his extended family had lost their houses in the Dixie Fire. He straddled his legs on the bike, sitting with arms folded across his chest. He'd been working as a hotshot earlier in the season before quitting for a better-paying job.

"I left. I regret it. I really regret it," he said.

"There's always next year," said Gene.

"You guys got the life though," he said, referring to the fact that they were employees of a private contractor. "Nice trucks. All by yourself. Getting paid way more than the Feds."

"That's true," said Gene with a chuckle.

"I would get my paychecks from the Feds and think I should be making at least triple this," the guy said. "I was gone two or three weeks at a time. I didn't have a life. It's a sacrifice.... I loved it, though. What's the biggest fire in California history—just over a million acres?"

"August Complex. This is the biggest single fire," said Gene, nodding to our surroundings.

"This town is already so fucked up now might as well let it be number one," said the guy on the dirt bike. "Fuck California. Let California burn. That bitch Gavin Newsom was in town the other day. I flipped him off out the window. He didn't look at me twice. It's never going to end."

"Nope," said Gene.

"Especially with our shit winter last year," he said.

"Someone posted a picture of Mt. Shasta yesterday," said Gene. "No snow at all. Guy said his grandpa is ninety-three years old, and it's the first time he's ever seen it with not a speck of snow."

"It just sucks. My beautiful valley. Born and raised. Just black all the way around it. It will be good deer hunting come August."

"Oh, yeah. The deer I've been seeing. Just crazy."

"You should have seen the buck I shot last year. I have a couple pets down here I like to hand feed."

"It will only get better with this drought because there's no water up high and everything is moving down low," said Gene.

"Archery season starts this Saturday. My biggest thing is I like to waterfowl hunt. Now I can't go... there's no fucking water."

"Do people talk about climate change around here?" I asked the dirt biker. "Is that on folks' minds?"

"There are very, very few Democrats around here. Most people are conservatives," he replied.

"I'm just curious how people make sense of all this stuff you're describing," I said.

"I mean, *I* don't believe in climate change," he said. "Yeah, a lot of things I say kind of lead up to it because everything is drier and drier every year. I don't want to believe it. It seems like it might be true, but I don't want to believe it. I'm a Christian. I'm a believer. This is like the end times in the Bible. I know where I'm going when the shit hits the fan."

Everything Imaginable Happened

WE WERE MOPPING up every day. I began to actually enjoy the tedium, the way time expanded and the hours spent hiking through the forest felt infinite. I studied the areas we hiked, trying to do forensics, looking at the way the fire had burned through the trees and why. The burn scars told a story.

We were in Bigfoot country, and I regularly wore my favorite baseball hat around fire camp, which said, "Teton Bigfoot Research Team." Some of the crew saw it as an invitation to tell stories of all the strange and bizarre things they had encountered in the forest: witchcraft, episodes of "lost time," Bigfoot signs. One of the crew members regularly captured photos of giant indentations in the dirt that they had come across during their peregrinations in the wildfire hinterlands.

I discovered that wildland firefighting revealed all of one's ignorance and weaknesses but also one's strengths. I could keep up on the hikes. I didn't "break line." I developed a sense for looking at soil composition and texture and guessing whether it was hot or cold. I could make a good cup trench, scooping the dirt with my rhino and pulling it upward with just the right amount of force into a gentle mound. It was during cup trenching that the crew seemed to have the most alternately dumb and profound conversations: Who was the best Winnie the Pooh character. Could boa constrictors slap you in the face. Fears of dating other wildland firefighters. Racism. Incarceration. Whether it was safe for gay wildland firefighters to come out to their peers. Heatstroke. Sometimes the shit-talking was so incessant that I found myself silently screaming in my head for everyone to just shut the fuck up.

We would return to fire camp in the dark, driving past outdoor stalls lit up in white lights and selling Dixie Fire T-shirts and sweatshirts. We picked up boxes of hot food from the canteen and ate dinner with our headlamps on, using the tailgates of the trucks as tables. Every night I struggled to decide whether to eat, shower, or sleep; there was never enough time to do all three. Sometimes I just rubbed my face with a wet wipe and crawled into my tent, leaving my throbbing feet out of the sleeping bag. But the temperature at night was dipping into the thirties, and I would wake up shivering and put on all my socks.

One afternoon, we began a mop-up patrol in a place called Lights Creek. It was clear that fire had raged through it. The trees were still standing, but many were charcoalized. The black stumps of willow bushes jutted out of the barren ground. As I walked toward the creek, my front leg sank to the knee in soft brown silt. Recalling stories of hidden ash pits and third-degree burns, I quickly pulled it out and stepped backward. With my rhino, I scooped out some dirt. It looked cold—but when I brushed my hand against it, I felt warmth.

"Hold for heat!" I yelled.

Another crewmate joined me, and we began to excavate. We dug through the powdery soil and sent up clouds of dust. The deeper we went, the hotter the ground became. The heat permeated the soles of our boots—eventually, we were dancing to relieve the discomfort. I stepped away from the pit and took in the situation. We were standing on an oven. Yards away from us, other crew members were also digging. Together, we were uncovering a single network of still-smoldering roots.

"We need water," the guy next to me said.

Someone radioed for an engine to bring hoses. A few people retreated to the rocky creek bank to wait, and I joined them. I leaned against my tool's wooden handle and drank through the tube of my HydraPak. I tried not to dwell on the bleakness of the scene—the stumps and leafless trees and cooked earth.

"Oh, shit," someone said.

I turned and saw a massive fire cloud rising over the mountain to the east. The smoke and vapor boiled and expanded. I took a

Everything Imaginable Happened 171

picture with my phone, then sat and stared. Over the next hour, I watched its white, cauliflower-like head rise. Later, I wrote to Michael Fromm, and he confirmed that what we had seen was a pyrocumulonimbus. During the hour we witnessed it, it rose to twenty-five thousand feet. I had only ever seen pictures of mushroom clouds before. I tried to imagine the combustion taking place below it—the heat and speed of a fire that could send so much smoke and ash into the sky. It was dreadful and awesome.

"How many megafires have you guys been on?" I asked a couple of the crew members nearby. We began to debate the definition of a megafire.

"OK," I said. "How many fires have you been on bigger than one hundred thousand acres?"

"I don't know if I've ever been on a fire *smaller* than that," said one.

"I would have to go home and count my sweatshirts," said the sawyer sitting next to him.

The water and hoses finally arrived in the evening. We sprayed the powdery soil, so hot that it hissed and screeched and steamed as the water hit it. When we finally drove out of the valley, the drainages to the south glowered with orange flames.

The Dixie Fire had grown by two hundred thousand acres since we'd arrived in Quincy a week before. From its origin in the Feather River Canyon, it exploded north, its flanks fanning to the east and west. Just before it had reached Lake Almanor, all the way back in July, its head had split into two, and now the polycephalous beast, driven by northerly winds, was wrapping back around itself. The entire Indian Valley had, in a clockwise direction, been encircled by flames, which were now backing down from the north into the Genesee Valley and threatening the town of Taylorsville. I had resigned myself to the idea that we were going to mop up for the whole roll. We were like sanitation workers cleaning up the fire's mess. I was disappointed, but then I thought, Be careful what you wish for.

One day, we drove up and down a single road pulling thousands of feet of hose that had been laid around houses. The hose would be moved to a flatbed truck and taken back to camp, rerolled, and sent to the nearby Genesee Valley. Pulling hose was a menial job. To do it, you had to "butterfly" each hundred-foot hose by extending your arms out on either side of your body, holding a nozzle in one hand. Then you had to bend your knees, squat low, and swoop your other arm to hook the hose, then swoop with the other arm, and repeat again and again until you reached the end. Jayson taught me how to twirl the butterflied hose off my arms when I was finished so that it came off already neatly tied, but no one could butterfly hose and look elegant. Everyone looked goofy, especially gargantuan Coby. I counted how many hoses Garrett carried across his shoulders to the truck, and then I tried to outdo him by one. The hose was abrasive, like a snakeskin made of flakes of glass. It rubbed the exposed skin between my wrists and gloves until it was raw.

But it was one of the best days. Everyone was in a good mood. The sky was relatively clear. The sun was shining. Even Jayson, whose tough veneer seemed impenetrable, appeared happy. He turned the music up full blast, careening around the turns in the road, shouting lyrics to Jelly Roll songs full of outlaw pathos. We pulled up in the driveway of an evacuated house. In the backyard was a Fleetwood RV, the same make used by the TV character Walter White as a mobile meth lab.

"Oh, man!" said someone. "You guys ever seen *Breaking Bad*?"

"Seen it?" Jayson growled. "I lived it."

THE GENESEE VALLEY was a sliver of flat land squeezed between two mountains with the Indian Creek winding down its middle. One of the mountains was Grizzly Ridge, where a spot fire had ignited and was now spreading and threatening the valley from the south. On his blog, Zeke Lunder described Grizzly Ridge as "some of the steepest, wildest ground in Northeastern California. I only know a couple of people who have even walked where the spot is burning." We were sent out one morning to improve a dozer line and prep it for firing operations that would prevent the spot from spreading to the south. We drove up and up on a dirt track to seven thousand feet elevation. It was a land of giant red firs. The trees had rumpled skeins of bark. Some of their trunks were four feet in diameter and splattered with neon-yellow wolf lichen. You could tell the snow line by where the wolf lichen grew; sometimes it was ten or twelve feet from the ground. We passed a mother black bear with two cubs playing with fir cones.

Our job was to clear all the small trees and shrub understory some ten to twenty feet in from the dozer track, and drag the debris to a chipper that moved alongside us. Everyone worked fast and intensely. There was no joking; we barely spoke. The saws were all running and screaming as they cut through the trees. The swampers were tearing at the dense underbrush and hurling things at us. The chipper whined and shrieked as it chewed through the conifer saplings we fed it. I wore my earplugs, ensconced in a bubble, and pushed and threw cut trees and logs down the slope to the road for hours. My nostrils were filled with gasoline fumes, sawdust particles, and the smell of Christmas. I felt tired in an entirely new way. I moved by sheer strength of will. I didn't know it at the time, but I was actually sick, my lungs riddled with the so-called camp crud, a kind of respiratory plague unique to wildland firefighters, caused by some toxic combination of dust, smoke exposure, dirt, dehydration, and fatigue.

The next day we headed back to Grizzly Ridge for another day of chipping. We staged in the forest, packing lunches and water and prepping the saws, only to be told to load up. New assignment. We were driving back down the mountain to the Genesee Valley.

Entering the valley, I had the impression it was surrounded by fire. Any direction I looked, I could see smoke tendrils in the forest. The caravan parked in the gravel driveway of a sprawling ranch house set amidst the trees at the bottom of Grizzly Ridge, which was now gushing grey smoke. A metal pole out front displayed the American flag. I followed Gene, the crew boss trainee, and Jayson over to another group of trucks emblazoned with the insignia of a hotshot crew. They were a Native American crew out of a southwestern reservation, and the superintendent leaned against the hood of his truck, scrutinizing an iPad; the screen showed a topo map of the ridges bordering the southern edge of the valley. His yellow Nomex shirt was grimy and ripped across the back of his shoulders, as if he'd been mauled by a tiger. It was nearly two p.m.—the witching hour, when the sun is hottest, humidity is low, and winds are strong. He zoomed the map in and out, then moved it left and right. He was trying to find a safe route for his crew to hike up and cut a line to slow the fire's approach. It was our crew's job to support them.

"It's a shit show," said the superintendent. "They want someone to get hurt."

"This don't look favorable," agreed Gene.

"When you really think about it, there's not much valley."

"You can almost let it do its thing and then start blacking it down," said Gene.

"You're a pyro, huh?" said the superintendent. "You're a pyro like me."

People coughed and kicked the dirt as radio chatter played in the background and the wind gusted.

"How would you attack it? I know locals know how to attack it," said the superintendent. Ten years ago, he went on, crews could engage directly—working close to the fire itself—because there was some moisture in the ground. "Now there's no rain. It's so dry," he said.

"Everything is available," said Gene.

"We can't go direct anymore, because it's going to stand up and chase you out."

"These winds start pushing through," said Gene. "This is kind of something they threw together too late."

"They could have had it two days ago," said the superintendent. "Two steps behind."

We parted ways to each scout the ridges for places where bulldozers and handlines could be put in to prevent the spot fire from reaching the valley. The four of us threw our packs and tools in the back of a truck and drove it up an old dirt mining road that followed Little Grizzly Creek through the canyon. As Gene looked at ridges to our right, trying to identify places where the fire could be cut off, he became even more skeptical about the plan. It was rough, steep terrain. As we turned around a bend, we saw a red engine parked on the side of the road. We passed it, and Gene did a double take. He asked Jayson to pull over, and we got out. Gene walked over to the engine and greeted a wildland firefighter with a white handlebar mustache.

"I was just telling her the story of when we were in Idaho and you came walking out of that Pioneer Fire," said Gene.

It was the ghost, the line scout who had appeared out of nowhere in Idaho and warned the crew to turn down their assignment.

A look of surprise came over the man's face. "Oh, you're *him*?" he exclaimed. "Right on! Remember that? 'Where's your fucking fusee? Get ready!'"

They laughed and shook hands. The scout told him he no longer worked for the Feds. "I'm getting too fat and sassy. I'm a private contractor. I *like* it. It's fun," he said. "That was awesome. I talk to these guys about the Pioneer all the time. Everything imaginable happened on that fire."

"Every day," said Gene.

"It's crazy," the line scout said. "You're the one guy that I know I can take over that ridge right now and know what's up."

"That day that burn operation got away from them...," said Gene.

"And we lived. It was pretty cool. I got some pretty cool footage. Don't worry about dying!" The scout laughed and then quieted. "That's exactly what I'm watching right now with this behavior. We were up on top—one twisted ankle and it's over. Everything about this thing screams out they're about to have a serious problem."

"We thought we could start pulling these ridges all the way to the top and punch in there," said Gene. "But looking at it right now, it's going to outflank us within the first couple hours."

"I'm insanely frustrated," said the line scout.

"We're still learning every day," said Gene.

They said their goodbyes, and we drove back down the road to the crew to watch as the fire crawled down the slope behind the ranch house, sending up streamers of white smoke. Individual trees began to flare. The smoke churned and became a luminous cloud, tinged peach on its edges. It shrouded the sun so thickly I could stare directly at the tiny, fuchsia dot it had become. I followed the crew boss trainee and Jayson into the woods to scout, climbing over fences and tracing the edge of the valley floor. Jayson found a baby rattlesnake and picked it up, holding its cheeks tenderly between his thumb and forefinger as its tail swished. We listened for the fire as we walked, tilting our heads toward the trees. We headed back to the trucks and the rest of the crew. A spot fire had erupted along the Indian Creek, and two K-MAX helicopters were already mounting an air attack, darting like dragonflies, dumping buckets of water to slow its spread. Using the creek as an anchor point, we were going to wrap the spot fire with handline.

"We've got spots everywhere," Gene warned us over the radio.

"Copy that: spots everywhere," someone repeated.

Frank turned to us. "Remember, guys, don't just run off to spots alone."

"Copy," we said in unison.

"Line it up!" shouted Jayson.

We assembled into a single file. "Moving!" we each shouted down the line.

We marched across the meadow, past a herd of black cattle huddled together that watched us closely, and counted off as we went.

"Fourteen!" I shouted.

Suddenly one of the squad bosses yelled from behind: "Hey! Maura! Coby! This way!" We split off and headed toward some flames burning through the grass, a six-foot spot fire ignited by an ember. We slung our tools, scraping a line at the fire's edge until it was circled, and then stamped on the smoldering grass to put it out.

The squad boss led us back toward the creek, where the crew was already working in dense stands of brush. Sawyers cut through the burning shrubs with wailing chainsaws, and the swampers shouted and grabbed the brush to hurl it away. The crew followed behind, clawing at the sandy creek bed with tools. The wind was gusting.

"That wind," I said to Coby. "Holy shit."

"Feels nice though!" he joked.

We got into line. The smoke heaved around us, and the helicopter rotors thumped overhead.

"Now you just push yourself to your fucking limit," said Coby. "Get it."

I put my head down and began digging line. I had discovered that if I switched dominant arms—six swings with my right and then six with my left—the muscles in my shoulders didn't cramp. I counted in my head, and sweat began to trickle down my back. I inhaled breaths slowly and always exhaled for a beat longer. When I felt the rotor wash of a helicopter overhead, I yelled, "BUCKET DROP!" and snatched a glance upward to make sure I wasn't directly under its waterfall.

Finally someone shouted, "Keep swinging till you tie it in! Ten fucking feet!" The line complete, we stood at the edge of the meadow to make sure it held. To the south, I could see fire heaving black smoke on the ground along the tree line we had been hiking. Some hotshots were burning out around the ranch house. I looked at Jayson. "Everywhere we were walking earlier—it's on fire," I said.

He looked at me with his eyes wide. "Uh... yeah!"

Gene laughed.

The wind began to shift to the south, reversing the flames and pushing them toward a copse of unburned willows. The crew boss trainee called in a dozer to cut a track along the fence line of the meadow and then turned to me. "It's a good thing we didn't engage today. This thing would have bit us in the ass," he said. "If we were up there"—he pointed toward the mining road—"this fire would have came around, and we would have been in a bad area, up canyon."

For the next few hours, we watched as the sky darkened and the fire made its way through the willows along the creek, sending up twenty-foot flames. We marched back across the meadow under an indigo sky, the hollows of the surrounding mountains glittering.

The next day we were back in the Genesee Valley. We walked into the mouth of the canyon where Little Grizzly Creek ran. The fire had spread all the way down the slope and was still hot. We cut line along its burning edge. The heat on our faces was intense, and we dug for hours, cutting through roots and dry duff that seemed bottomless. One of the crew tried to extinguish flames crawling up a tree trunk and got a radiant burn that flushed her cheek a deep scarlet red. When we finished the line, we began laying hose and pressure-washing the ground, sending up squelching bouts of steam from the soil.

In the afternoon, I walked to the end of the line, where it met the shallow, pebble-strewn creek, and walked across. Further up the creek, the fire was still moving through the forest. The smoke was diffused through the trees, veiling them in white haze. Ethereal flakes of ash floated through the air. I sat still on the embankment, an uneaten ham sandwich spread on my knee while I watched a black and white butterfly drink the sweat on my hand with its proboscis. Nearby, Frank sat leaning against his fire pack with a blue pack of American Spirits in hand, smoking.

Frank was shy. But it felt like he had taken me under his wing, quietly giving advice, pointing to places I should avoid, making sure I had a lunch in my pack before we left camp every morning. In turn, I always made sure to check the ground where Frank rested or worked for the gloves he often forgot, and gave him ibuprofen and electrolyte tablets from my pack. I didn't know how, but Frank seemed to outpace guys young enough to be his kids. He seemed to relish sacrificing his body, volunteering to dig or chase spots even ten hours after we had first started out and the light of day was fading away. I had gathered little pieces of his story. He cursed in anger any time a PG&E truck drove past. The Camp Fire, which had been started by a downed power line, had wiped out his hometown of Paradise and burned his house down. The investigation as to what started the Dixie Fire wouldn't be finalized for months, but it was already clear the company's power lines were likely responsible—again.

Now I asked him: Was it difficult to be on the Dixie Fire considering what had happened to him on the Camp Fire? "It's not more difficult. It just pisses me off. The way it all went down with PG&E

and the whole fucking thing," he said. "Arrogant bastards. They don't even give a shit." He was hoping for a payout from the Fire Victim Trust, the settlement fund set up by PG&E, but he didn't expect to get enough to rebuild his home. "As a company, they don't care anymore how many fires they start, how many people they kill," he said. "They're pretty much getting away with murder. They admitted to it, and nothing happens to them. Why?"

As he smoked, Frank spoke in a halting and looping manner, taking long pauses when he reached a difficult topic, as though the silences explained more than he could with words. He had grown up in San Jose, California. His paternal grandparents were immigrants from Japan. During World War II they had been sent to an internment camp, where his dad had been born. "They were living in concentration camps for *hella* years," he said. "They had to make their own pots and pans and furniture, everything." During the summers, Frank went to Nogales, Mexico, to stay with his mother's family. He loved music and played with a punk band, rollerblading from San Jose to San Francisco along the highway just to get to practice. He struggled with substance abuse. "I had to grow up real quick, real early. I was pretty much an alcoholic when I was eight or nine. Going to school, wanting a beer," he said. "I remember thinking I was never going to make it past eighteen. You know what I mean?"

Frank had spent almost a decade in and out of prison as an adult. "I wasn't the best person for a long time," he said. When he was in his forties, he decided he wanted to be a better father to his kids. He accepted a prison term knowing it would give him a chance to get into a wildland-firefighting camp. He saw it as a path to rehabilitation and a job. After his release, he applied to the private contractor four times before finally getting hired. That was five years earlier. "I'm still a drug addict even though I don't put anything in my system," he said. Wildland firefighting gave him structure, a mission. "My past... trying to be way better," he attempted to explain. "I get enjoyment helping these guys out. How can you not feel free out here? It's a great feeling. It's not for everyone.... I love it."

On the morning of November 8, 2018, Frank had been heading to work in Chico when he saw a plume of smoke in the sky from the direction of Paradise. He ran into Dan Kelleher at a gas station, and they discussed the new start. "It didn't look to me like it was where they said it was," explained Frank. He headed to pick up his crew. "I turned on the scanner, and I heard it was right there in the canyon. I said, 'Screw this.' I called my wife to tell her to get the fuck out of there. I knew it was really close, and my mom was there, she's handicapped. My stepson was there." Frank had a friend drive him to his house, and he gathered everyone, five people and a dog, into his vehicle to escape. "We drove out, and it took us four and a half hours."

"Do you feel because of your training you understood what was happening at a level others didn't?" I asked him.

"Absolutely," he said. "Absolutely. Who knew it was going to go off like that? Only the people who knew about fire. That was the craziest thing I've ever been in. It was really weird. It shouldn't have happened that way. Everything was perfect, I guess. Way perfect."

"Were you scared?"

"I was worried about getting them out. They were freaking out. I'm scared of fire but not to the point that... I don't know. It was just burning way way way way hot. I was watching the road catch on fire. There wasn't even fire around. It was just the heat in the air just lit the streets up right in front of me. Pavement. Yeah. I never seen that."

"What happened after?"

"I went back the next day," he said. His house was gone, but the thing that bothered him the most, he explained, was losing the watches his grandfather had made from pieces of scrap during his years in the internment camp.

"It sucks that you can't pass something down to somebody," said Frank. Two days after he escaped the Camp Fire, Frank was back working on the same fire.

"What kind of work were you doing?"

"We were cutting line," he said. "I got pulled off after a week to do a chipper on search and rescue. Clear everything away so that they or I can go in and look. That was weird, too. What we were looking for... having dogs find them."

"Also, because you must have known people, right?" I asked.

"I believe I knew some of them. Not even five hundred feet from my house was a retirement home. The whole thing got wiped out. They were just old people you see at the store and start talking to 'em."

"I imagine that losing your house was difficult."

"At the time, it wasn't difficult. I got everybody out. That's a whole other story—the aftermath. It's been two years, and we can't do anything with it. I bought a trailer and live on the property. I'm gonna rebuild there and hopefully use it as a rental. I'm not going to stay there anymore. We're going to get out." He paused. "I've seen a lot of crap. All those things in the Bible that nobody thought would ever happen, I've seen those things happen."

I left Frank and walked back down the fireline where a couple of the sawyers were resting next to an enormous fallen tree trunk. They offered me some mint-flavored snuff. "Just like gum," they said. I put some on my hand and inhaled it. I was so sick by then, the buzz barely registered. I lay down on the massive log and stared at the sky. Twenty minutes later, I suddenly woke up. Everyone around me was still talking and cracking jokes. "I've never seen anyone do snuff and fall asleep," one said.

I went back to the end of the line and found Jayson preparing to scout. By now, I trusted Jayson so completely I would have followed him anywhere. We walked through the woods quietly, stopping every now and then to stand and listen and look up or through the trees for signs of wind and flame and embers. We could hear the fire somewhere back there, a soft roar like the ocean in a seashell. As we walked, a hotshot crew emerged in front of us in a single-file procession. They looked like a fey band of Lost Boys. Their faces were grim, smeared in soot, their hair matted after tussling with the beast in the forest. Their tools hung like totems over their shoulders. I wanted to call their mothers. *Don't worry, I saw him.* We moved aside until they passed.

By eight o'clock that night, our crew was hiking out along the line back to the trucks. Engines were showing up for the overnight shift. Their emergency lights flashed like strobes, but the emergency was everywhere all at once; there wasn't anything to do about it. Most people just stood and watched as the slope of mountain directly to our north began to ignite, the flames moving westward like a slow tsunami. I stood in the meadow. The ground of the forest sparkled with incandescent heat illuminating the trunks and crowns of trees in black silhouette until one by one they each caught fire and their crowns flared, some so brightly it looked like lightning in reverse, a crack of white heat sent from earth into sky. These were Norman Maclean's "giant candles burning for the dead," I thought. Soon the whole of the mountain seemed cast in molten magma, and by the time we left we drove through a forest glowing sublimely, so much fire everywhere that no line could ever contain it and all there was left to do was sleep.

8   CIRCLING THE SQUARE

THERE HAVE BEEN THREE KINDS OF FIRE ON EARTH, ACCORDING TO THE HISTORIAN Stephen J. Pyne. The first kind was natural fire, which began with the evolution of plants hundreds of millions of years ago and the lightning that ignited them. The second was human fire, when the last glacial surge receded and our species domesticated fire in hearths and on torches and spread flames over the landscape like little *Homo arsonists*. "Fire and people forged a mutual assistance pact," said Pyne. He believes this pact constituted a new geologic epoch that he calls the Pyrocene. The third kind of fire is unprecedented. It is combustion "no longer bounded by such ecological limits as fuels, season, sun or rhythms of wetting and drying," according to Pyne. Instead, we have entered a "pyric transition" in which we removed open flame from our daily lives and exhumed the lithic landscape to fuel chemical combustion in special chambers on an industrial scale. The by-products of this new kind of fire have led to changes in the atmosphere and overwhelmed the planet's capacity to cope.

Our current stage of the Pyrocene is defined by too much combustion. Humanity has engaged in a binge of burning in which there is a lot of bad fire and not enough good fire. It is a time of the "dominion of fire.... A radical reconstitution of climate as fire makes a world more favorable to fire." The Pyrocene, Pyne says, appears "so dire, and its likely trajectory so strange, that some observers argue that the past has become irrelevant."

If we wanted to see what the future of the Pyrocene might look like, the Dixie Fire—with its firestorms, military checkpoints, refugees, burned towns, and futile containment strategies—is as good an example as any. Air tankers dumped flame retardant worth over $200 million on the fire. Seventy-eight helicopters and seventy-seven airplanes attacked it. Eighty-nine handcrews and 569 engines were mobilized. It cost $610 million to fight the Dixie Fire. Still, it burned for more than a hundred days.

The day after I left fire camp to return home, the rest of the crew returned to the top of Grizzly Ridge. The spot fire had grown and was threatening to move toward a major highway and the town of Quincy. The crew assisted several hotshot crews by cutting line and tying it in with bulldozer tracks for a planned burnout operation. Some of them thought it was risky; the weather reports weren't good. But they were instructed to light a fire off the line, so they did. For two days, my phone lit up with photos of the fire burning amidst the giant firs we had worked underneath the week before. After the crew was demobilized and went home, however, the operation went awry: fire spotted across the dozer line, and intense winds pushed it to run east. "It still took off just like we said it would," Jayson texted. The firing operation ended up increasing the size of the Dixie Fire by another hundred square miles.

Zeke Lunder continued scouring maps every day and posting analysis to his blog, *The Lookout*. He could see on the infrared images how crews lost control of their burnout operations again and again. It wasn't surprising. A good firing operation ideally involves weeks of preparation, he pointed out. On the Dixie Fire, operations were being carried out with just a few days of planning. The real surprise was that high-risk firing operations were still being executed under terrible conditions in the first place. Zeke figured crews just didn't have any other option. Burnout operations, however futile, were one of the few tools available to them. He wrote, "A tougher question for us to face might be, Why are we still focused on containment, when it's clear that parts of this fire are beyond our control?"

Garrett Rangel liked to say that wildland firefighting was harder than it looked but a lot less noble than people thought. "Just a bunch of fire pirates out here," he said. A nihilist at heart, Garrett viewed wildland firefighting as a mostly destructive act. I knew he wasn't wrong. Firefighting riddled the forest with dozer tracks, painted it red with retardant, set fires to contain others, deepened the fire paradox. The Dixie Fire was a case study in the excesses of what some called the fire-industrial complex and the futility of suppression. The Dixie Fire showed the dominion of fire.

Actually, Pyne disagreed with those who thought that in the Pyrocene, the past was irrelevant. On the contrary, he believed that surviving the epoch demanded that we recover the accumulated wisdom of our species' relationship to fire. We needed to bring back forgotten kinds of fire.

WHEN I RETURNED to California in early September, the Dixie Fire was still burning. But I wasn't going back to the crew. Instead, I drove east from San Francisco into the state's parched, flat interior, past the homogenous almond plantations with dark understories and into the hills clothed in gold. Off the main highway, on roads that bent and curved, I rolled down the windows to let in wafts of sun-baked pepperwood trees and grass, the scents of my adolescence. When I arrived in the town of Mariposa it was twilight, but I kept driving, following the two-lane highway north until I found a campsite for the night.

I had returned from the Dixie Fire in miserable physical condition. I lost my voice and spoke in a whisper; the camp crud had stripped my vocal cords like turpentine on paint. It took a week for the smell of smoke to leave my hair; my legs were bruised; the skin on the soles of my feet eventually began to peel off. I was also bizarrely euphoric. For weeks, I experienced an inner quietue bordering on peace. It was as though the labor and discipline of firefighting, the total immersion in the materiality of trucks, tools, wood, dirt, and fire, had stilled some thrumming tension in my core. Had I really felt such a desperate need for utility before? The political philosopher and motorcycle mechanic Matthew B. Crawford has said that when we become competent in a practice, it changes our perception. Our attention becomes constrained, and the world takes on a definite shape. This "ecology of attention" lets us reclaim the real and become a "powerful, independent mind working at full song." Full song, full song, I now hummed to myself.

In the morning, I drove back to Mariposa and met Steve Pyne in the parking lot of his motel. Steve was in California to conduct research at Yosemite National Park. I had been in the park only once before—as a kid with my grandpa—but I knew it was one of Dan Kelleher's old haunts. In the mid-aughts, he was spending half the year at Yosemite conducting prescribed burns. For a few days, I would follow Steve around as he observed fire scars, giant-sequoia groves, geography, and geology, putting the pieces of the park's fire history together to write a treatise on the topic. Steve now explained we would focus on Tioga Road, which traversed several fire scars, including that of the Rim Fire, which burned a quarter of a million acres in 2013. The park, Steve pointed out, had been in part protected from the Rim Fire's worst effects thanks to its decades-old program of prescribed burning. Then we would visit two of the park's three groves of giant sequoias because, as he put it, "that's where the park actually began; that's why it was set aside."

Now in his seventies, Pyne had short white hair that was covered with a green baseball cap, and a pen was visible in the front pocket of his blue button-down shirt. He was relaxed and cheerful, eagerly prepared for a day in the field. We got into his silver rental car and chatted about Mariposa. I noticed right away that Steve suffered from the affliction common to wildland firefighters: he could not look at a place without imagining how it would burn. Steve's diagnosis of the flammability of the old mining town, which was aligned on a north-south axis between two ridges, was not good. "It's a chimney chute," he said, shaking his head. The previous night, I told him, I had been woken by the faint sound of raindrops pattering my tent, the result of dry thunderstorm cells pushing northward over the area. They had also illuminated the sky with thousands of lightning strikes. Now we wondered if the strikes had started any new fires in Yosemite and whether we would get a chance to see them.

Steve had a hypothesis about Yosemite. He saw it as a place in which the past and present of fire were condensed, and where its future might be gleaned. A kind of microcosm of Earth, Yosemite had been shaped by ice and then fire, he explained. The Ahwahnechee had practiced setting fires here for centuries before rangers and park superintendents violently forced their removal in the nineteenth century. The fire exclusion that ensued had weakened and threatened the survival of the park's ecosystems. In the 1960s, a revolution in ecology led to efforts to reintroduce fire and pay down Yosemite's fire deficit. One focus of these efforts was the Illilouette Creek Basin. Located in the backcountry of the park, the Illilouette was not a popular tourist destination like Half Dome. But since 1972, it had been the focus of a progressive experiment in which lightning-ignited fires were carefully allowed to burn rather than be put out. Over the next fifty years, fires equal in size to seventy-five football fields had burned through the sixty-square-mile area. The result was a remarkably complex landscape: mature trees were widely spaced from one another while younger trees regenerated in patches, and dead, burned tree snags dotted the landscape. Unlike other parts of the drought-stricken park, the basin had maintained its grasslands, shrublands, and wet meadows. Water in the basin was actually retained with greater efficiently because gaps in the tree canopy allowed more snow and rainfall to reach the ground, resulting in reduced competition between trees. Unlike the forest that surrounded the basin, the Illilouette had been made resilient.

Steve was planning a three-day trek with pack mules into the Illilouette alongside a dozen of the park's managers and rangers, as well as Scott Stephens, the professor at the University of California, Berkeley, who had studied the impacts of fire policy in the Illilouette for twenty years. Steve thought the success of the Illilouette Basin was a potential beacon of good fire policy for the rest of the country. Maybe it could lead us through the coming ecological darkness. "Yosemite is one of the most visually iconic places in the world," he said as we drove. "It has all these *genius loci*, from John Muir to Ansel Adams. Yet all the major decision-makers are going to this valley, this shallow valley, that nobody knows anything about, that has never been celebrated or photographed. Why? Because, in some ways, it is the future."

JUST A FEW days earlier, Pyne's latest book, *The Pyrocene: How We Created an Age of Fire and What Happens Next,* had been published by University of California Press. Publication days were no longer extraordinary events for Steve. How many books had he written? By my count some twenty-eight over his career, plus a nine-volume history focused on the legacy of fire in every region in North America. He had been thinking and writing about fire for most of his lifetime and had toured the world, from Russia to Canada to Mexico, scouring archives—at times teaching himself new languages in order to read original texts—in a quest to understand the element and its relationship to ecology, history, and culture. Steve described his scholarly style as "omnivorous." He wanted to know everything possible about a subject in order to understand how it all fit together.

The origins of Steve's interest in fire weren't abstract or even academic. His preoccupation with the subject stemmed from a career as a wildland firefighter. A few days after he graduated high school, Steve signed up to work on a fire crew on the North Rim of the Grand Canyon and returned every season for the next fifteen years to chase smoke. It was, he wrote, "good, tough work." He became a foreman for the crew, composed of an amalgam of college students, tribal members, general misfits, and hell-raisers. They called themselves the Longshots. "For three or four months," he wrote in his book *Fire on the Rim*:

> we will live, work and talk together and fraternize with almost no one else. We are infatuated with our camaraderie, and it is exclusive to the North Rim. The North Rim is our world, and its concerns—and its concerns only—are ours. Within limits, our identity is ours to make. We dispatch ourselves. We outfit ourselves. We train ourselves. We name ourselves. From the Park there is little leadership. What really defines us is fire. Fires solve all problems: fires make everything possible.

Steve viewed his seasons smoke chasing and firefighting as a sort of rite of passage during which he transitioned from adolescence to adulthood. In that same time period, he graduated from Stanford, got married (to one of the Longshots), received a PhD in American civilization (taking his doctoral orals with his fire boots on), and welcomed his first daughter. Had it not been for bad knees, back, and wrists, he might have continued chasing smoke forever. Writing about fire was a way of maintaining a proximity to fire even as his body failed him. While working as a wildland firefighter, however, Steve had begun to wonder about the implications of national fire policy and the intellectual ideas that shaped it, as well as the historical tensions between fire as a threat and as a necessary ecological process. "The problem is to reconcile one hybrid of nature and culture, fire, with another hybrid, wilderness," he wrote in *Fire on the Rim*:

> The philosophical complexities are bottomless. They are soon matched by the difficulty of trying to translate what is fundamentally a national creation myth—wilderness—into a program of field projects. From the mid-1970s to the mid-1980s most of the disastrous wildland fires in America were the result of breakdowns in prescribed fire programs. It will take decades to learn the techniques. The philosophical issues, as John Dewey suggested in another matter, will never be resolved, just "gotten over."

Steve's intellectual passions were not widely shared. He struggled at times to find support for his research, bouncing between grants and fellowships, including a MacArthur Fellowship—known as a "genius grant." He found it difficult to establish a permanent foothold in academia. Fire as a subject, Steve has pointed out, lost the pedigree it once shared with the other elements centuries ago. "The other elements have academic disciplines behind them," he liked to say. "The only fire department on a university campus is the one that sends emergency vehicles when an alarm sounds." Fire was relegated to a subset of chemistry and thermodynamics or applied fields like forestry and "began a long recession from the life of the mind." Steve, however, remained convinced that it was important for humans to maintain their imaginative capacity for fire. He saw our historical amnesia about the marvelous relationship we once had to this element, our lost understanding of its necessity to ecological health, as one of the causes of the era of pyrrhic loss and disaster we now found ourselves in.

In the summer of 2021, as the Dixie Fire neared gigafire status, Steve's convictions that fire was worthy of our intellectual consideration, and that it was critical we give it that consideration, seemed vindicated. The more we sought to make fire disappear, the worse the fires got. For the preceding few years, Steve had watched megafires burn from the window of his home in a suburb of Phoenix, something he could barely have imagined at the start of his career on the North Rim. In *The Pyrocene*, Steve wrote that the term "pyrocene" "has the advantage of showing how today's binge-burning happened, of reminding us that we had a fire crisis before climate change threw accelerant onto the flames, that fire will not go away when the climate stabilizes or reverts to earlier forms, that fire is what we as a species do." Our future would abound with flame: either we would ignite the fires to prevent the bad, or the bad would proliferate.

WE DROVE THROUGH the Stanislaus National Forest on a road that scrawled along the bottom of the dizzyingly steep Merced River Gorge, a notoriously fire-prone place. A few years earlier, the Ferguson Fire had blown through it, burning nearly a hundred thousand acres and taking the lives of two firefighters. Driving with Steve was like being treated to a master class on the history of science, geology, two centuries of American conservation, and National Park Service folklore. It turned out that Steve's mother, who had just celebrated her ninety-ninth birthday, was born in a campground in Sequoia and Kings Canyon National Park and spent her first night in a suitcase. He had visited Yosemite as a child and even seen "Firefall." When Steve described this event to me, it sounded so outrageously risky I hardly believed it was real until I saw archival footage of it myself. Beginning in the 1870s, every summer evening, Yosemite's concessioners started a bonfire at the top of Glacier Point and let it burn into embers, then pushed the coals off the edge of the cliff to fall down the three-thousand-foot rock face. The glowing embers created a cascade of light as they dropped to the valley floor below. In 1968, the pyrotechnic display was finally quashed when it was deemed unnatural. "They were right to finally stop it," said Steve. Still, he admitted, it had been spectacular.

At Arch Rock Entrance, Steve flashed a permit allowing him to enter the park. Driving along the floor of Yosemite Valley, I caught flashes of soaring granite faces and rock spires between the trees. Yosemite Falls came into view, but it had become a dry fall weeks earlier; all that was visible was the discolored stone, a specter of water. We noticed that some of the trees in the valley were charred black six or seven feet from their bases, evidence that a prescribed fire had recently burned there. "Look at those flame lengths," Steve marveled. We pulled into the parking lot behind the Yosemite Visitor Center and walked to a cluster of wooden staff buildings painted the uniform brown of the Park Service. At a picnic table under some towering incense cedars sagging with a crop of gold-frosted cones, we met a man in green fire pants and a navy-blue T-shirt emblazoned with the insignia of the Yosemite Fire program. His wavy brown hair was streaked with grey at the ears, and he smiled at us amiably while carrying a large cylinder of rolled-up papers under his arm. His name was Dan Buckley, and he was the branch chief of fire and aviation for the park.

Buckley unrolled the papers and spread them out on the picnic table. Each one was a map of Yosemite's past fires, some going all the way back to 1930. The map he focused on, however, showed all the fires burning at that very moment in Yosemite. "This one is still active," he said casually, pointing to the symbol of a diamond located in the Aspen Valley. "It's still growing. It's twenty-three hundred acres now and grew about three hundred acres last week. It's headed into the rocks. We went out and put line around the bottom of it to just kind of corral it so it would run up a canyon into some rocks. Nobody's out there dealing with it, so we're not wasting resources." We listened, quietly impressed. Buckley continued pointing to diamonds on the map. "The Lukens is still putting up smoke. County Line is still putting up smoke. There's two fires up here, the Paiute and the Cold. Cold is kind of interesting because it's right off the Pacific Crest Trail. So, lots of calls from hikers. We had to put up signs that say, 'Don't call us.' "

For each of these wildfires, Buckley's crews had taken different strategies based on the topography, weather, fuels, and ecological benefit. They didn't snuff fires out; they herded, nudged, pinned, encouraged, or simply watched fire "do its thing." When satisfied with their efforts, they left the wildfires to move through the landscape for weeks. "Way back, the philosophy was you had to have a good reason to put a fire out," said Buckley. "We're trying to reinstill that philosophy here. Why are we putting a fire out instead of letting Mother Nature take its course?"

The 2020 Blue Jay Fire had burned for *six months* before it finally extinguished itself at seven thousand acres. "The park is 747,000 acres," explained Buckley. "We should be burning about 16,000 acres a year, on average. That's the historical average looking back at the fire record before white settlers came in here. That's probably what we need to maintain it. We don't come close to that unless we get one of these large fires."

This approach to wildfire sometimes collided with public expectations. "There's been a shift from Smokey Bear—fire is bad—to the need for prescribed fire," said Buckley. "But if you're here on vacation and Yosemite's been on your bucket list your whole life, and you can't see Half Dome because of smoke, I get it. I say, 'Sorry we ruined your vacation, but what would you have done if it was foggy and rainy?' Smoke is just as much a part of the environment. I can't blame myself that God made the woods flammable."

"This has been happening as long as plants colonized continents," agreed Steve.

"That's why we burn it under our conditions and not when it's hot, dry, and windy," said Buckley. "We're trying to get it burned so it's not nuking everything."

The previous night's lightning storm had ignited three new fires in the park that they knew about; Buckley suspected there could be more. As we loaded into his truck, he rolled the windows down so we could see clearly and smell the air for wafts of smoke. "You guys are kind of fire lookouts, so if you see something tell me," he said. "We don't know what we have, and that's usually the case for two or three days." We drove north along Big Oak Flat Road and then Tioga Road until Buckley pulled the truck onto the shoulder to point at something across a broad, grassy valley.

"You can barely see the tuft of smoke coming up above the ridge there. If you look right over this tree and kind of to the left side of the tree, you can see a little bit of black smoke," said Buckley. I looked and saw nothing.

"You can see that, Steve?" I asked.

"If I'm seeing it correctly, behind the top of the tree, there's a rounded mountaintop and to the left there's a depression," Steve replied. "Right in there." The start was so new it didn't have a name. I stared in vain. Steve's fifteen seasons as a smoke chaser gave him special powers, I decided.

"To the left of the tree is a ridge, and you see a kind of building?" said Buckley. "Come down that ridge about halfway down.

That's where Brian Hughes, Arrowhead Hotshot captain, died on the Ferguson Fire."

Hughes had been part of a felling operation on a steep slope and was crushed by a ponderosa pine when it fell downhill instead of up, a full 145 degrees from its intended lay. The tree was fifty-seven inches in diameter and over a hundred feet tall.

"We hiked out there on the second anniversary of Brian's death," said Buckley. "It's so steep. Took us almost an hour to go half a mile."

Buckley had been an Arrowhead Hotshot for sixteen seasons, but he hadn't intended to be a firefighter. Three days after he graduated from high school, he had gone to Sequoia and Kings Canyon to work a summer job at a concession stand. But he started hanging with the fire crew on his days off, and they showed him the ropes of wildland firefighting. Eventually he joined the hotshots.

We ate lunch in an empty campground called White Wolf. Through the trees, Buckley pointed to a few puffs of smoke near the ground. It was the Lukens Fire, still burning weeks after it had first ignited, benignly creeping around. We finished eating and drove a short way up the road to Crane Flat, the site of one of the oldest helitack crews in the country. The helicopters were landing and taking off, doing reconnaissance on the new lightning starts. A couple of Type 2 handcrews and a wildland fire module—a ten-person crew specialized in fire-behavior analysis and managing natural ignitions—were on their way to help manage the fires. So-called fire modules had been pioneered in Yosemite, Buckley told us. They were designed to be highly mobile and self-sufficient, capable of spending days isolated in the backcountry observing and manipulating wildfires or doing prescribed burns. "They tie in with natural barriers, do a little handline, fire it out, call it good," said Buckley. There were now around fifty wildland fire modules in the country, and the one arriving at the park that day had a female captain. "It's great to see people rising up in the ranks and pushing us old white males out of the way," he said, "which probably needs to happen."

"Not right away," I joked.

"Well," he said, "pretty quick."

IF YOU COULD pinpoint the year that the fire revolution in America started, it was probably 1962. That March, a group of otherwise upright and respectable citizens gathered at Florida State University in Tallahassee to do something crazy and backward: stand in public and advocate for igniting fires intentionally. The event was called the Tall Timbers Fire Ecology Conference, and some of the presenters were nervous: fire-lighting was still looked down on as destructive, irrational, even sinister. They had been "damned uphill and down," as one put it, for their heretical opinions. But the annual conference became the epicenter of what Steve called in his book *Between Two Fires: A Fire History of Contemporary America* a "fire counterculture." A belief in the ecological benefit of fire gained followers and began to impact fire policy in national parks, forests, and on private land.

In April of that same year, a botanist and ecologist for the conservation organization the Nature Conservancy lit some matches, "tossed them into dry bluestem," and torched twenty acres of prairie in Minnesota. The "April burn, done wholly with volunteers outfitted with rakes, matches, and metal backpack pumps, was a bold bid," Steve wrote. "A science of fire ecology did not exist by name. It was tricky enough for fire civilians to burn autumn leaves, much less open landscapes designated as critical habitats."

In July of 1962, the First World Congress on National Parks was held in Seattle, and the findings of the "Leopold Report," a document written by Starker Leopold, a professor at the University of California, Berkeley, and the son of the naturalist Aldo Leopold, were formally accepted. The report delivered a radical message: America's parks could no longer be roped off and passively expected to thrive. Elk herds were exploding, other species were declining, and forests were becoming overgrown, vegetative tangles. In order to create the conditions and "biotic associations" that "prevailed when the area was first visited by the white man," wrote Leopold and his coauthors, active interventions were needed, including the restoration of fire as a natural process.

Soon after, a young graduate student by the name of Richard Hartesveldt published a dissertation on Yosemite's Mariposa Grove that would become the basis of a significant report titled *The Giant Sequoias of the Sierra Nevada*. The trees are some of the most fire-adapted species on earth. A single tree produces hundreds of thousands of seeds each year, but none will sprout into a sapling until fire has heated the tree's cones to disperse them. Only when a seed has landed on a bed of soil cleared by heat and flame and made fertile with ash can it begin to grow. Furthermore, giant sequoia saplings need a forest free from shade-tolerant competitors and an open canopy created by fire. The report found that protection from fire was devastating giant-sequoia groves because the trees could not reproduce themselves. A couple of years later, Harold Biswell, at the University of California, Berkeley, began burning on Redwood Mountain in Sequoia and Kings Canyon National Park, which contained the continent's largest sequoia grove. In the years that followed, some of California's giant-sequoia groves became the site of bold experiments in prescribed fire. "Those were the good old days," someone who participated in them told me. "You didn't have to have a red card; they would just shove a drip torch in your hand and say, 'Go light the forest on fire.'"

The heyday of the fire revolution was arguably the 1970s. After that, it suffered one setback after another. During the Reagan years, according to Steve, environmental agendas stalled. Urban sprawl into rural areas (what some called "living in the stupid zone") demanded increased fire protection. Wildfires themselves began to erupt in size and intensity, and drought afflicted western states. When over five million acres burned in 1988, including in Yellowstone (where some of the fires started as a prescribed natural fires), suppression again became the dominant response. "By the end of the 1980s virtually every index showed that there was significantly less fire on the land than a century before," Steve wrote in *Between Two Fires*. He called the 1980s the "lost decade," in which the fire deficit ballooned and biotas—the plant and animal life specific to certain regions—crumbled. Then the 2000 Cerro Grande Fire in New Mexico, the result of a prescribed fire that spread out of control, soured the public on intentional ignitions. (The Cerro Grande didn't, however, deter Buckley, who afterward tried, and succeeded, to be the first person in the western national parks to undertake a controlled burn.)

There were pockets where fire continued to flourish. The southeast. Gila National Forest. Kansas's Flint Hills. And Yosemite, where the fire program continued under Jan van Wagtendonk, the park's research scientist and a former doctoral student of the pioneering fire practitioner Harold Biswell. It was van Wagtendonk who developed a concept of "prescribed natural fires." Rather than suppressing every wildfire start, he wrote prescriptions for the conditions under which wildfires should be allowed to burn in order to maintain "wilderness character." The idea was an attempt to circle a square, in Steve's opinion. It mediated and reconciled two separate constructs: people and wilderness.

When Buckley came to Yosemite, one of the first things he did was make an appointment with van Wagtendonk, who had been at the park for almost three decades by then.

"I'm the new prescribed-fire and fuels specialist for Yosemite," Buckley told him. "What is my job?"

"Just put fire on the landscape," van Wagtendonk said. "Let it do its thing." Buckley called him "the master."

Buckley left Yosemite in 2004. After van Wagtendonk retired in 2009, drought, big wildfires, a staffing purge, and waning urgency gutted the park's mission to bring fire back to the landscape. It was always going to be easier to take fire out than to put it back in, and this was true across the country. And now, sixty years after the start of the fire revolution, America still lacked the capacity and willingness to reintroduce fire on a large scale. There was no national workforce mandated to put fire back by planning, preparing, igniting, monitoring, and mopping up prescribed burns.

Jennifer Mueller of the nonprofit organization the Ember Alliance told me that today we urgently need an "army of fire practitioners." Creating one would require another reimagining of fire's place in American life—like the one that occurred in 1962. "By that I mean planning for wildfires, doing community protection, cooperative burning with federal and nonfederal sources, creating good, blue-collar jobs at scale," said Mueller.

A good place to start would be to make basic firefighting classes more accessible, explained Mueller, by creating teaching materials that include women, people of color, underrepresented groups, economically poor communities, and young people. The fire world should reflect the demographics of the United States. "Primarily, [wildland firefighting] is white dudes," she said. "There are starting to be more women, but it really doesn't reflect the diversity of our country, not even close." Mueller herself had been training as a biologist when she witnessed her first prescribed fire. "I was like, 'Oh, this is *it*.' I was hooked. I didn't even know that prescribed fire was a thing, much less that *I* could do that. If you don't see somebody who looks like you doing the job, you don't necessarily think that you belong."

The irony is that there is already an existing workforce of skilled fire practitioners. But, as Jeremy Bailey, the director of prescribed-fire training at the Nature Conservancy pointed out, their time and resources are almost entirely dedicated to emergency fire suppression. For this reason, Bailey sees the wildfire problem as a workforce organization problem. If, at a federal and state level, the same funding and prioritization were given to starting fires as they are to suppressing them, the fire deficit would soon disappear. "I can easily imagine the difference we could make if we were able to focus thousands of wildland fire practitioners on controlled burning over a six-month season," Bailey wrote for a blog post on the Fire Adapted Communities Learning Network. "Imagine if for every firefighter poised and ready to extinguish any start, we also had a fire lighter."

ON OUR LAST day in the park together, Steve and I hiked a couple of miles along the Washburn Trail toward the park's most famous giant sequoias, the Mariposa Grove. We shared the trail with tourists eager to see the iconic trees, but as we entered the grove, the crowd seemed to disperse and we were left alone. We passed the gargantuan trunks of a giant sequoia lying horizontally on the ground. It was one of the trees that had blown over the previous January when a Mono wind—cold air that is funneled down the western slopes of the Sierra Nevada at speeds of up to a hundred miles per hour—had slammed into the park. Altogether, fifteen giant sequoias and hundreds of other trees had fallen. The Mono wind had started the first fire of the year; a week later the park was under four feet of snow. "Things are moving much more quickly," said Buckley of the bizarre sequence of weather. "Storms are much more energetic. I liken it to putting a pot of water on the stove and cranking up the heat. Things start moving around, and that's what we're seeing with our atmosphere."

Muir had called groves of giant sequoias "sublime wilderness." He described how "the trees with rosy, glowing countenances seemed to be hushed and thoughtful, as if waiting in conscious religious dependence on the sun, and one naturally walked softly and awestricken among them." We were awestricken now as we approached the Grizzly Giant. The massive, rippled trunk was so large and incongruent with the surrounding trees, so exceptional, that my instinct was to doubt my eyes rather than believe its reality. The tree was eighteen hundred years old. According to park lore, more than fifty billion people had been born and died during its lifetime. I stood at its base, which measured thirty feet in diameter, and looked upward until, as I tried to glimpse its crown, I could not bend my neck anymore. I curved my back to strive to see its top, straining to comprehend. Steve, his backpack slung over a shoulder, stood in front of the tree's cat face—a blackened fire scar, over twenty feet tall. How many fires had it experienced? Hundreds, no doubt. Then Steve also tilted his head upward, directing his vision into the tree's enormous crown. A single branch of the Grizzly Giant exceeded the size of nearly any other tree species. Up there were dozens of bird cavities and as many as two billion leaves supporting lichens, insects, and chickarees. It was like staring into another galaxy. "If you had to describe it," he said, "it would hardly fit the definition of a tree."

For Steve, fire was sui generis, the beginning and end of the story. He always started with the question Why is there fire here at all? This single query led him to geology, ecology, chemistry, culture, art, myth, colonialism, and politics. "There's lots of themes, history, people, and stuff that fire can illustrate in interesting ways," he told me. "No, for me, fire is the organizing principle. The rest of it falls around that." He had often been alone in this inquiry.

When his first fire book, *Fire in America,* was published in 1982, one academic told him he had treated his subject in a kind of epic and exalted way, but if fire was really that important, somebody important would have already written about it. "A part of me could be despondent from time to time. Why don't people see this? Why don't people recognize there is a bigger story here?" he told me. "I was like someone walking along the American River outside Sutter's Fort in 1848, just picking up these nuggets. Why aren't there half a dozen people like me writing about this? It was just out there for the taking. If I had another twenty years and support, I could keep going where no one else has gone." Steve accepted that people might never find fire as interesting as he did. But now every fire season demanded our attention. "It's not that I was clairvoyant and knew what was coming," he said. "I wasn't. It's just that the world changed."

As we left the grove, we talked about the landscape around us. Had it been created by nature or culture? "It's the wrong question because it's not an either/or," explained Steve. "I think we've figured out that it was the result of an interaction." The future terms of this interaction had now become critical to the survival of Yosemite and the other forests. When Steve first visited Yosemite as a child in 1957, the atmosphere's carbon dioxide was 315 parts per million. Sixty-four years later it was 420 parts per million. "The solution is a process. There is no end point," he said as our feet thumped downhill. "Everything is changing; we're on a changing planet. We have to be able to pass between two fires. We have to be able to live with fire. You start simple, fail, and build out. As you change things, you interact with what's changing. You have to be skilled, willing to learn from and question what you're doing and just keep at it. The persistence is really important. If you fall behind, then things get really impossible. We're falling behind every year. Burning is maintenance. You have to do this forever."

In addition to managed wildfire, Steve believed that prescribed burning had to expand to a scale not yet seen. We needed fires that roamed across landscapes over longer burning seasons. Our decisions about when and where to light needed to be informed by science as well as by folk wisdom and traditions. He had invented terms to try to capture this approach: Loose-herding. Burning-out wildfire. Hybrid fires. Free burns. Fire-foraging. One of his favorite examples of this strategy, he told me, took place in 2014 on San Carlos Apache Reservation in Arizona. In April of that year, dry lightning had started two fires, one in grass and shrubs and the other around a creek. They became the Skunk Fire, and rather than attack it, the reservation's fire management team decided to construct a box around it by piecing together roads, ridges, trails, and hand and dozer lines. In an essay he wrote about the fire, Steve pointed out that historically, the Apache had not burned as meticulously or extensively as other North American tribes, but this was mainly because they lived in a place where fires were already so frequent. "Their primary contribution to the region's fire regimes was not to add ignition so much as remove barriers to fire's free propagation," he wrote. Before the Apache were forced onto reservations in the late nineteenth century, fire had been free-ranging, and that was what the tribe now sought to do with the Skunk Fire—allow it free range. Over several weeks, the San Carlos fire crews herded the Skunk Fire using observation, burnout operations, retardant, and aerial ignitions, ensuring it went where they wanted and stayed within the box they had assembled. When it finally died out, an area twenty-two miles long and eight miles wide had been treated with good fire.

"Prescribed fire's got to be more mobile," said Steve as we walked. "It's got to be much larger. Just have people follow the snow. Follow the changing conditions as things become ready to burn. Crews would have a very large area so they're just out there, not necessarily monitoring, not fighting. It's a different kind of engagement. Fire is not just a tool; it's a relationship that's constantly adjusting."

I kept thinking about the Lukens Fire. It had started in late June when lighting ignited nineteen new fires in the park. Each one was assessed based on its location and potential impacts. For the Lukens, the wildland fire module decided to use two roads as a southern perimeter and then lit backing fires near them, ensuring that the fire could only progress to the north toward a burn scar. By the time we saw it, the Lukens Fire had burned 867 acres over a time period longer than that of the Dixie Fire, but it was so benign we literally had a picnic next to it. It was feral... but also intentional. It was the circled square.

The narrative arc of the story Steve had discovered on the North Rim of the Grand Canyon as a smoke chaser—maybe even earlier, as a child in Yosemite watching Firefall—was becoming clearer to him. As we reached the end of the road, he said, "We're recovering our history. Fire is a part of our heritage, and we're restoring it to what it was like. It's a redemption story."

9 RESIST, ACCEPT, OR DIRECT

CALIFORNIA'S GIANT-SEQUOIA GROVES WERE THE SPIRITUAL SPARK OF THE CONservation movement and later became the focus of some of the most progressive fire experiments in the America. Yet until recently, an estimated 80 percent of the state's groves lacked fire for over a century, and the iconic trees are more vulnerable than ever. In 2020, the Castle Fire, a high-intensity megafire, torched the crowns of hundreds of giant sequoias, killing an estimated 10 to 14 percent of the total world population. Fears that another disaster could endanger the species' survival are high, and the trees have become a fulcrum of renewed debate over how we should protect them and the country's forests in the age of the Pyrocene. Do we have an obligation to intervene and restore forest health? What should such interventions look like? Should they include drip torches? Chainsaws? Logging trucks?

In Yosemite, I saw firsthand why these questions have become so urgent. The park's giant-sequoia groves were surrounded by thick, overgrown forests prone to hot and fast wildfires. Walking with Steve Pyne along the Washburn Trail to the Mariposa Grove, I realized I, too, had fallen prey to the firefighter's predilection: looking around, all I could see was the blowup that might engulf the place with just one spark. Just a few feet off the trail, all was darkness and entropy. The packed trees blocked the sunlight with their spindly, dead branches that overlapped one another. It was a tinderbox. Even the smallest ground fire would easily climb into the crowns and run through this place, I thought. When we walked off the trail to eat lunch on a log, the dead branches snapped like bones under our feet. I put my hand on the ground and began to rummage through the duff. It was powder dry. I dug and dug and never got to the bottom of it.

The Tuolumne and Merced Groves weren't even surrounded; they were enmeshed in a forbidding tangle of drought-weakened fuels. One day we parked at a dirt trailhead to hike to the Merced Grove, descending downhill through overgrown ponderosa pine, incense cedar, and white fir along what used to be an old stagecoach track. The grove was the smallest and most secluded of Yosemite's three groves, made up of just fifty mature trees. On the way, we encountered one of the park's botanists, off-duty, with his toddler hoisted on his back. "I wouldn't want to be trapped down here in a fire," he said, his eyes wide with alarm at the density of the forest around us. The botanist described the fuel-treatment proposal he was developing to try to protect the groves by re-creating presettlement tree density: cutting down hazard trees and stems less than twenty inches in diameter throughout two thousand acres of the park, and hauling the biomass to a nearby mill or processing plant. As far as Dan Buckley was concerned, the proposal was critical for protecting the sequoias. "It's called maintenance," he said. "If you don't do your maintenance, things start to fall apart, the ecosystem falls apart." We didn't know it at the time, but forces were already gathering to kill this project. Environmental groups claimed it was commercial logging in disguise and filed a lawsuit. Thinning trees, the lawsuit argued, would fail to conserve Yosemite's scenery in a way that would leave it "unimpaired for the enjoyment of future generations."

Meanwhile, the lightning storm that swept over the Sierra Nevada while I slept in my tent had ignited two fires in Sequoia and Kings Canyon National Park. From a containment perspective, the starts were in particularly bad spots: the terrain was steep and rugged, making it difficult to get people or equipment like engines and dozers in. The park, like the rest of the state, was in the midst of a historic drought, and the forest was choked with a lot of dead and downed fuel. The fires spread quickly and grew into one another, becoming the KNP Complex. It was, people predicted, going to go big.

By that September, the nation's fire preparedness level—a ranking of fuel and weather conditions, fire activity, and fire suppression resource availability—was at its highest possible point. There were multiple complex wildfires across the West, and at least 80 percent of wildland firefighters were already mobilized on other fires. Sequoia and Kings Canyon National Park was struggling to even get an incident command team assigned to the KNP Complex. Buckley was in the parking lot of a grocery store when he got a call from Sequoia and Kings. Could he send some people from his park down to help them out? "It's like if we had multiple wars all over the planet," Buckley told me. "And we have another war break out, so we send a platoon of marines to fight another country."

The chance of containing the wildfire with handcrews and tankers was pretty slim, so the park focused on protecting nearby communities and the giant-sequoia groves. Wildland firefighters wrapped the base of the 275-foot-tall General Sherman Tree, the world's largest tree by volume, in silver aluminum material that shrouded its thirty-six-foot-diameter trunk from intense heat. The image of the ancient behemoth in a dendrological fire shelter whipped around social media and elicited profound dread and grief from the public. The fire was growing by thousands of acres each day, and it started making runs into some of the groves. On September 17, 2021, a head fire ran into the Suwanee Grove, resulting in moderate- and high-severity burns, and then spread into western areas of the Giant Forest Grove. The next day, it pushed further east across Generals Highway and threw a spot fire just four hundred feet from General Sherman. It seemed like the KNP Complex was becoming a worst-case scenario.

Some staffers at Sequoia and Kings Canyon National Park, however, were wondering if the wildfire was a boon in disguise for giant sequoias. One of them was Christy Brigham, the park's chief of resources management and science. As the KNP burned through some eighty-six thousand acres and sixteen groves over the next month, she took note of the fire's behavior. There had been a few head fires, but most of the groves experienced backing fires or burned under a thick inversion, which quieted the fire's intensity. To Brigham, the groves were desperate for fire treatments anyway. She started thinking that maybe the fire was going to have way more upside than downside.

When the KNP Complex was fully contained in mid-October, Brigham started doing a burn-severity analysis. She found that altogether, 2,589 acres of the groves experienced low- to moderate-severity fires, and 616 underwent high-severity fires. By her estimate, between 1,330 and 2,380 sequoias over four feet in diameter were either killed or would die in the coming three to five years. Not great news. The Windy Fire, which burned to the south of the KNP Complex around the same time, had also resulted in giant-sequoia mortality, meaning that some 3 to 5 percent of the Sierra Nevada sequoia population was lost in 2021. But it wasn't the unmitigated disaster that people had feared. And the wildfire had opened up new opportunities for protecting giant-sequoia groves in the future. In places that had received previous prescribed-burn treatments and mechanical thinning—decreasing tree density with machines and chainsaws—severity was reduced, and crews could implement safer strategies for suppression operations. Overall, the KNP Complex had given giant sequoias potentially important benefits: reducing fuels, creating small canopy openings for regeneration, removing litter, and creating ash that would nurture the next generation of seedlings. Some studies had even shown that more severely burned groves might benefit too: twenty-five years after some local groves had burned in moderate- to high-severity wildfires, the density of stand regrowth was abundant.

When I talked to her on the phone, Brigham explained how the "first entry" of a fire into an ecosystem that hasn't seen fire in a century is the most difficult, dangerous, expensive, and risky. The KNP Complex had, in a sense, done the hardest work for them. "Now we have all this newly burned landscape that generates opportunities," she said. "We can use that burn as an anchor to do other prescribed burns." Although the first entry can kill a lot of trees, it doesn't get rid of all the fuel; hence the need for more burning. The KNP Complex had created a five- to ten-year window in which officials could utilize the black it left behind and create another reentry of fire and remove the trees that had died. It might take yet another reentry after that to fully restore the landscape's fire resilience. But before they could do any more treatments in the groves, a litany of legally required processes would have to be carried out: planning, compliance, public notification, species surveys, archeological surveys. Brigham worried that these would drag on until the park missed the very opportunity the wildfire had handed them. "The existing law and policy around both environmental compliance and air quality and historical resources were all built sort of in the sixties and early seventies with the idea that the actions we take are bad for the environment," she told me. "They're set up to make sure that when we take an action, we fully evaluate its consequences with the idea that our action is bad and the worst-case scenario. That's totally not true now. The wildfire that we're trying to prevent is the terrible action."

THE SPEED AND intensity of the Castle and Windy Fires had alarmed those who study giant sequoias. "The effects of these fires were just off the chart," said the dendrochronologist Tom Swetnam. "It's an irreplaceable loss when you lose a two-thousand-year-old tree. On a really long timescale, the mark of something extraordinary happening is when we lose a grove entirely. And that's extremely possible. The groves themselves are separated, and some of them are just ten isolated trees, and ten miles away is another little group."

Tom had been studying giant sequoias since the 1980s. At that time, he had just concluded his PhD research on fire scars in tree rings in the Gila National Forest, where some of the country's early experiments in prescribed natural fires were taking place. During Tom's second year with Gila's helitack crew, one of his jobs was to ride horseback with a pack mule and follow the flames of the Langstroth Fire for weeks as they crept around, eventually burning over three thousand acres before the fire was snuffed out by autumn weather.

It was a controversy that first piqued Tom's interest in giant sequoias. During a prescribed burn in Sequoia and King Canyon's Giant Forest Grove, a crew member purposefully created a hot fire in order to build a convective column that would pull embers upward and away from nearby sequoia logs. The fire scorched the bowls of some trees, and although the burns were found to be superficial and tree growth within the grove increased, the images of the revered giants with great streaks of blackened bark led to newspaper articles and a public outcry, including criticism of the park service's burn program by environmentalists. The park decided that a rigorous scientific study of the species' fire history was needed to justify reintroducing fire to the groves. Tom contacted Jan van Wagtendonk at Yosemite and expressed his interest in undertaking the study. "I was lucky," Tom recalled. "We did eight years, going back every summer, and sampled six groves."

Tom told me this story in his backyard office up in the Jemez Mountains, where he grew up and where he had returned after his retirement from the Tree-Ring Lab. He swiveled in his chair to find a framed photograph behind him. It showed a verdant forest thick with ferns. A young Tom stood on an enormous stump of a sequoia that had died and fallen, its exposed root bulb dwarfing him. He wore an orange hardhat and was bent over the tree's colossal trunk, his forearms bulging with effort as he held a chainsaw with an eighty-inch bar to cut a cross section. The scale of human to tree was so improbable that I was once again struck by the likeness of a giant sequoia to a whale; it looked like Tom might have been harvesting the fat off a humpback. Sometimes, he told me, they used a six- or eight-foot bar to get radial sections from the trees. They had to make vertical cuts because if they held the saw horizontally, the chain was so heavy it would simply fall off. Each cross section of tree he cut was the size of a dining table and weighed hundreds of pounds. The samples were so wet, it soaked their hands to touch them. "We rented a thirty-foot U-Haul truck every summer and filled it with samples," he said. "By volume, it's the single largest collection at the [Tree-Ring Lab's] archive."

The feeling of working in the groves was one of privilege and awe, even fear. "I would call it a power spot. There's something that is glowing there," he explained. "They're special places, and there's no denying it. You can see it. We were using these cutting tools—not in the living ones, the dead ones—but even then, we were defacing them. Sometimes I wondered: is this the right thing? I felt like I was in a sacred, holy place."

The findings from Tom's eight-year study of giant sequoias appeared on the cover of the November 1993 issue of *Science* with a close-up of a tree's rings singed with fire scars. With the samples he took, he was able to create a two-thousand-year fire history of the groves. It showed that the fire regime was highly variable, depending on the climate. During warmer periods fires were frequent, and during cooler periods they were less so. But the longer the interval, the bigger and more intense the fires were when they finally ignited. This was because of the buildup of fuel; fires led to less fuel and smaller, patchier subsequent fires. Tom described the size and frequency of fire in the groves as in constant flux. "This long-term view supports an ecological paradigm that emphasizes the ubiquity of change in ecosystems, rather than tendencies toward stasis or climax communities," he wrote.

Tom believed that the evidence of frequent fire during periods of higher temperatures and drought indicated that the rate of prescribed burning needed to dramatically accelerate in the era of global warming. More fire was the best way to prepare the trees for the climactic conditions to come. "The giant sequoias have survived warm periods in the past, and they did so with the highest frequencies of fire ever recorded," he told me. For forty years, he had been pushing for aggressive action. But it never happened at the pace and scale that he believed was needed. "There's still a fire deficit," he said. "There's still too much fuel in most of the groves."

MANY OF AMERICA'S forests are arguably in a condition similar to or worse than their state in the 1960s, when Starker Leopold warned in his seminal report of an unprecedented accumulation of impenetrable vegetative tangles on the nation's lands. "Much of the west slope [of the Sierra]," he wrote, "is a dog-hair thicket of young pines, white fir, incense-cedar, and mature brush." Leopold explicitly warned of the danger to giant sequoias and described the problem as a "direct function of overprotection from natural ground fires." He advised reintroducing more fire in order to maintain a state similar to the one that existed when European Americans first arrived.

One of the people who fully embraced this message was Nate Stephenson, an ecologist with the US Geological Survey who helped with the creation of Giant Sequoia National Monument. Nate had collaborated with Tom Swetnam in studying sequoia groves and became an expert in the species' ecology. By the 1990s, however, the idea that parks could be returned to a precolonization state began to crumble for him. "I let go of Leopold's vision kicking and screaming. What triggered it was the parks' inability to come even close to burning enough each year to match pre-Euro-American fire regimes," he said. "When my hope collapsed, they were at maybe 15 percent of what needed to be burned. It threw me—not an exaggeration—into a depression about the park service and its future."

This was around the same time that the Intergovernmental Panel on Climate Change issued one of its first reports showing that humans were influencing the climate through the burning of fossil fuels and the buildup of carbon dioxide in the atmosphere. Nate saw climate change as a potential runaway disaster barreling down on the country's ecosystems, including giant-sequoia groves. Then, in 2014, what had previously been chronic but subtle background changes seemed to suddenly become massive and acute disruptions. "That was the year we started to see hotter drought effects on giant sequoias that no one had ever seen before," he recalled. "[In] 2015, we had our first really nasty wildfire that burned a lot of sequoia groves and killed a lot of big trees." The Rough Fire killed a hundred trees, which at the time was unimaginable; prior to that, the most that had been destroyed by fire in a single year was fourteen. In subsequent years, the Railroad Fire and the Pier Fire killed another hundred trees or more. For the first time, Nate witnessed big sequoias torch in their crowns and die. "That caught us off guard," he remembered. Then came the Castle Fire in 2020. "Not only did sequoias torch; there were crown fires, and I don't think anyone knows, but [there were] maybe even running crown fires that were independent of ground fire." Nate had conducted the Castle Fire study that estimated that 10 to 14 percent of the world's population of giant sequoias had died. "From the perspective of someone like me, who has worked among the sequoias for so long, it's like, 'Oh my god, it's falling apart before my eyes.'"

Nate told me he saw three options for responding to climate change: resist, accept, or direct. He saw resisting as unrealistic and accepting as morally untenable. Directing, in his opinion, required thinning the forests that had become overgrown from fire suppression and weakened by pests, disease, and drought. Even the Leopold Report had hinted at the necessity of such a dramatic step: it said that chainsaws and bulldozers might be required in order to reintroduce fire. "Trees and mature brush may have to be cut, piled, and burned before a creeping ground fire can be risked," Leopold wrote. Some fire ecologists call this restoration process "reintroducing disturbance," and it is not easy to do. As the plant scientist M. Kat Anderson has explained, fire behaves differently than it did a century ago because the structure, type, and amount of fuels are so different from the historical baselines. Sometimes an intermediate step like thinning is required, followed by burning, and then sometimes reseeding, removing invasive species, and stabilizing soil. "Restoration is a long-term process," she wrote in *Tending the Wild*. "Over the long term, a restored ecosystem's integrity will depend on reestablishing the mutually beneficial ecological associations between it and people."

Scientific studies have consistently shown that the structure of North American forests has undergone massive change over the last century. In one recent peer-reviewed paper, Tom Swetnam and thirty coauthors argued that overwhelming evidence showed that predominantly open-canopy forests once dominated by mature and old trees have transitioned into denser forests with smaller trees. The result is forests that are more fire prone than at any point since European colonization. What is needed, they contended, is a "paradigm shift" to depart from fire suppression toward supporting fire-adapted forests and communities through active forest restoration that includes thinning trees mechanically.

But cutting down trees is highly controversial. For decades, North American forests were savaged by extensive logging that destroyed most of their old-growth stands. Environmentalists who fought to end these practices resist the idea of putting machines in wilderness areas and fear that such strategies would be exploited by federal agencies and timber companies. Some environmentalists have even disputed the need for less-dense forests, claiming that the harm of megafires on landscapes has been overblown and that their high severity is consistent with historical fire regimes. They advocate for defending homes and communities but leaving the forest entirely alone—neither intentionally igniting fires nor putting them out. Some have even argued that thinning trees makes wildfires worse by reducing the forest's capacity to be its own windbreak and by opening up the canopy, which creates hotter and drier conditions. For years, activists used federal protections like the Endangered Species Act to obstruct or end wildfire prevention projects.

One such environmental group is the John Muir Project, a California nonprofit that has repeatedly litigated to stop forest-thinning projects, including those proposed around giant-sequoia groves in Yosemite. According to the organization, the US Forest Service is a governmentally sanctioned logging corporation that has created a myth of the need for forest management in order to mask timber grabs. In one lawsuit against a Forest Service plan to remove hazardous trees along roads in Sequoia National Forest, the director of the John Muir Project wrote, "I enjoy being in forests in a natural state, and a logged environment ruins that enjoyment."

In the fall of 2021, the *Sacramento Bee* published an extensive article about the ongoing debate between the scientific community and some of the environmentalists who are against forest thinning. In the preceding months, Tom Swetnam and over a hundred other scientists had coauthored some forty-one articles rebutting the views of environmental activists and the methodological biases and errors in the arguments against forest management. "Wildfire Experts Escalate Fight over Saving California Forests," read the newspaper headline. In the article, the fire scientist Crystal Kolden said the very idea that "we shouldn't manage anything because the forest takes care of itself is a completely racist and very, you know, colonial viewpoint that ignores the thousands of years of extensive indigenous landscape management across California and the West." Some compared the environmental activists to early climate change deniers, who created an illusion of scientific debate around an issue on which there was clear consensus.

It seemed to me that the fight over forest management had become an emotional lightning rod exactly because it struck at the heart of a cherished American ideal: untouched wilderness. Challenging this myth by pointing to the historical and ecological evidence of human management, or suggesting that "wilderness" was a culturally specific idea born in an era of violent Indigenous erasure, elicited intense resentment. When the *Los Angeles Times* published an op-ed calling for collaboration between forest managers and Indigenous fire practitioners, the head of an environmental institute in California wrote a heated response. "The call by non-Indigenous people to incorporate a romanticized, stereotyped version of Native

American fire use is a thinly veiled attempt to appropriate Native culture for the same reason colonial powers have done so in the past—for self-interest or financial gain," he wrote. Forest managers were "demonizing nature" by turning "rich, dense vegetation" from "habitat to 'fuel' that must be removed."

Tom told me that he believed a contributing factor in this fight was a lack of trust in the motivations of ecologists who favor forest restoration. " 'Timber interests control you; you're going to make it worse," he summarized. "So leave it the fuck alone, and do the wildfire protection work around the neighborhood." But Tom only had to look out his office window into the Jemez Mountains to see a landscape that had been ingeniously managed for centuries. "They're carefully dancing around it but not really grappling with the fact that there are big chunks of landscape that *were* managed by people for long periods of time. It's antithetical to their primary ideology, which is rooted in the idea of not mucking around with the forest," he said. "They don't want to accept that it is possible for human beings to sustainably manage forests."

There was a radical middle between environmentalists and timber interests in the debate over forest thinning, according to Nate Stephenson. "I call it the timber wars of the southern Sierra Nevada," he said. "On the one hand you had companies wanting free rein to do a lot of logging, and on the other extreme you had environmentalists who didn't want humans to touch forests at all." He spent years of his career urging timber companies to change their practices from clear-cutting to small-group-selection harvesting. And, to environmentalists, he had a warning: "If you don't do anything, you're gonna lose it. It'll burn up."

NINE THOUSAND FEET above sea level, standing on the slope of Humphreys Peak, the tallest mountain in Arizona and part of the San

Francisco Peaks considered sacred by many Native tribes, I faced due west. As I hopped around in a cold autumn wind, I watched the sun set over what seemed like a stunning vista: a carpet of green trees expanding for hundreds of miles in front of me. This forest stretched from the south rim of the Grand Canyon across the Mogollon Rim to the White Mountains in the east. It was the largest continuous ponderosa pine forest in the country. But where I saw majesty, the man standing next to me saw an aberration: an overgrown, drought-stressed, pest-ridden landscape that had been neglected for generations. "Don't even look at it," he half-joked. "It's shit."

His name was Neil Chapman, and he was the forest health supervisor for the Wildland Fire Management program within the Flagstaff Fire Department. A diehard advocate of prescribed fire, Neil had spent his adult life trying to protect and restore fire-adapted landscapes. One of his favorite quotes was from Aldo Leopold, who had written in *A Sand County Almanac*, "An ecologist must either harden his shell and make believe that the consequences of science are none of his business, or he must be the doctor who sees the marks of death in a community that believes itself well and does not want to be told otherwise." His initiation in fire had taken place during his early twenties, burning on the Nachusa Grasslands in Illinois. Since 2006, he had lived in Arizona, where his work involved developing new strategies to implement mechanical timber harvesting. "In this part of the world," he explained, "wildfire risk reduction is all about getting trees out of the woods, putting them on trucks, putting them in piles to try and create conditions for good fire across the landscape." Neil had gathered a group of wildland firefighters from South Dakota, Colorado, Nevada, and southern Arizona for a training exchange event, or TREX, that would last a couple of weeks. I was in Flagstaff to be a part of the crew and to see how a forest could be restored to something like a precolonization state using a combination of modern machinery and fire.

Flagstaff is an urban wildfire zone. Situated in the middle of the Coconino National Forest, 99 percent of the houses in the county by certain estimates were at risk if a wildfire blew through. The city is a classic example of what the wildfire community calls the WUI, the wildland-urban interface. In the 1980s, the areas around Flagstaff regularly began experiencing fires larger than ten thousand acres. It became clear that the forest was suffering a death by a thousand cuts: historic buildups of fuel, longer fire seasons, and a changing climate. "Part of us was like, that's what we want, right? We've been missing [fire] for a long time," said Amy Waltz, a research scientist in Flagstaff. "But burn severity was also increasing, and our historical forests didn't used to burn in high severity."

Flagstaff was in a race with what locals referred to as the "Big One," a single, large-scale fire that could sweep through the city of sixty thousand and burn it to the ground. Many expected that it was just a matter of time before the Big One came. Faced with this existential threat, the city took out a $10 million bond to do forest restoration: thinning ten thousand acres of trees every year for ten years, and introducing fire back into the managed landscapes. Environmentalists criticized the concept as a "new silvicultural mythology," but it had found surprising purchase amongst the citizenry: 74 percent of voters had approved the bond. Flagstaff's model of forest restoration was now going to be implemented on a much larger national scale with billions of dollars of federal funding: the bipartisan Infrastructure Investment and Jobs Act of 2021 had earmarked over $3 billion for hazardous fuels treatment, and the 2022 climate and tax deal was slated to put another $1.8 billion toward thinning some fifty million acres of forest around vulnerable communities.

But in order to restore forests, you have to understand what they once looked like. What is the historical range of variation for different attributes in the forest? What did the forest look like before industrial-scale logging and fire suppression? What is the prescription?

In northern Arizona, the vast majority of logging had taken place relatively recently, between 1880 and 1930, when roughly 80 percent of the merchantable volume of pine was cut and sold. Due to the relatively recent logging history and the southwestern climate, the questions about what "restoration" means can be answered by looking at the record of the forest that is still in the ground. To re-create this record, scientists at the Ecological Restoration Institute, at Northern Arizona University, searched for old-growth stands as well as the detritus of the logging empire: tree stumps or even dips in the ground where stumps once stood. From this evidence they created maps capturing the stand structure of the forest that existed a few hundred years ago. The results were startling. In 1876 there were an estimated 23 trees per acre in the forest; in 1992 there were 1,253 trees per acre. In the intervening decades, the ratio of trees to open grassland was inverted, from around 81 percent grass openings and 19 percent tree canopy, to 93 percent tree canopy and just 7 percent grass openings. Historical ponderosa pine forests had complex spatial structures that included big individual trees, some trees in small groups of varying ages, and openings without any trees at all.

"Restoration is a holistic concept," said Dave Huffman, director of ecology at the institute. "We're not just focused on fuel buildup or changes in fire characteristics or patterns; we're focused on all those other attributes that we know are important: the understory, the forest floor, the soil, the wildlife habitat."

OUR CREW'S PLAN was to burn around eight hundred acres of ponderosa pine forest, some of it located within Flagstaff city limits. But rain had made things too wet to burn and delayed our operations.

As we waited for the ground to dry, we headed into the forest to use wedge prisms—pocket-sized tools for estimating the basal area of trees, a common way to describe a stand's density—conduct wildfire-risk assessments on homes, and learn about mechanical thinning.

One morning we set out in a caravan of pickups and engines and drove to an undeveloped area on the north side of the city. It had undergone aggressive restoration to try to approximate the presettlement conditions of the ponderosa pine forest. We parked on a dirt road and walked into what felt like a sun-drenched, grassy meadow dotted here and there with smatterings of pines. Standing in the grass was a silviculturalist—someone who plants and harvests trees to attain specific forest objectives—who had written the prescription for this land as part of an initiative called the Four Forest Restoration Initiative. The project encompassed 2.4 million acres in the Apache-Sitgreaves, Coconino, Kaibab, and Tonto National Forests. Its goal was to restore historical fire regimes and tree density across eight hundred thousand acres.

In the silviculturalist's hand was a hundred-page document that he called his "bible for restoration." It contained guidance on everything from managing ecosystems for Mexican spotted owl and northern goshawks to how to make decisions on keeping or cutting large trees. "I sat about half a mile over that way, reading through this document," he told us. "In the prescription I wrote, 'We wanted to open up the forest to mimic the spatial patterns that existed on this landscape prior to European settlement.'" The big yellow-bellied pines—those over 150 years old—would be retained, but the logger would punch holes amidst the stands of younger trees to open up the canopy. Oddly, the smaller trees could almost all be dated back to a specific year: 1919. That was when a massive cone drop rained down on the bare soil created by intensive logging, sheep grazing, and fire suppression. "All the regeneration that would have normally been killed in two to twelve years by a fire, survived and recruited," he said, meaning the young trees became a part of the overall population. He pointed to a nearby stand. "These trees with the black bark, the 'black-jacks,' are all from that 1919 seed drop."

The result of the restoration project was around forty to sixty trees per acre in a seemingly random pattern. Small groups of trees were left with interlocking crowns to support squirrels, one of the main prey species of the northern goshawk. Otherwise, the groups of trees were separated from one another with wide open spaces. Regarding the success of the initiative, "A lot of it has been building trust," said the silviculturalist. "In the early 2000s, there was, like, no trust in what the Forest Service was doing. We've definitely built up more trust that we're doing the right thing, and we're getting the flexibility to take the densities down more. We're getting on the same page when it comes to cutting big trees—we're still not cutting old trees—but big trees to create these openings between groups."

In fact, the true historical levels for tree density in the forest around Flagstaff were even *fewer* trees than we were seeing— somewhere between ten and forty trees per acre. But, as Chapman explained, any time they reduced the tree count to that density, there was an inevitable public backlash. "We have stakeholders at the table who for many years spent millions of dollars to shut down the logging industry [from] cutting big trees," said Chapman. "Now we're saying we need to revive that industry to cut down small and medium trees. We don't have the social license to get to that lower historical range, because some of those groups, I think, feel that if we do that, we'll need to cut some bigger trees. Or the forest just seems a little too open to them. In my mind, I don't really care if you think it's too open or not; this is what the science tells me. But I'm not going to go to war over it either."

The place we stood in was a compromise or, as Neil preferred to call it, "restoration with a dose of reality." He was happy as long as the wildfire risk was reduced enough so that there were still trees to argue about.

ON MY LAST day in Flagstaff I got into a truck with two wildland firefighters from South Dakota. They were unlikely best friends. Eric O'Connor was a short, bagpipe-playing, churchgoing white guy. He joked that his hands "make a Whopper look huge." Justin Bauer was over six feet tall, a member of the Fort Peck Tribes, and a Hunkpapa and Assiniboine descendent of the Lakota and Nakota people. He was also a combat veteran who spent four years in Iraq and Afghanistan and had forearm tattoos, a thick black beard, and a passion for rodeos, guns, and horses. They had met running a wildland fire crew together in Rapid City and were now on a multiweek, three-thousand-mile road trip together. After Arizona, they were heading to Oregon to complete advanced training in felling trees. I squeezed in the backseat beside shopping bags full of white bread they had brought from South Dakota, and we set out.

Our route took us north through the Kaibab National Forest and into Grand Canyon National Park. As we followed the road east along the South Rim, we began to see haze in the air and eventually dozens of piles of cut trees burning. We pulled into a parking lot, where staff from the park's fire program were waiting. One of them was a wildland firefighter dressed in a dirty yellow and a fire pack; he had just come from burning the piles. His name was Jason Nez, and he was an archeologist, a resource advisor on wildfire operations, and a member of the Navajo tribe.

"Do you want to talk about the real history?" Jason asked us. He wondered if we'd noticed the gas station just outside the park featuring a diorama of a big stagecoach and a bunch of Indians holding it up. I had. The fake diorama and crude mannequins were so huge it was almost impossible not to.

"Man, there's no recorded documentation of any Natives holding up a stagecoach here," he said. In fact, the park's existence had depended on the forced removal of his and other Native tribes. He then cited just a few of the numerous documented incidents of Indigenous displacement that had taken place. In 1897, for instance, the Coconino County Sheriff formed a posse and removed Navajos from the east end of what was then Grand Canyon National Forest Preserve. In 1934, park officials burned down traditional Havasupai homes and forced the residents to pay rent to live in cabins instead. What was important to remember, Jason said, was that no one alive today was present for those events. "We weren't here. We weren't part of that decision-making. We can avoid those things and find common ground."

For Jason, finding common ground started with acknowledging the park's creation myth, which said that because no one lived there, the Grand Canyon had great wilderness values. "We want to believe everything is fine," he said. "We want to believe that it was noble. But the issues I deal with and that we all deal with when we start things with mismanagement and miseducation and injustice—it tends to come back in bad ways. When you start something based on a lie, when you start something based on a falsehood, it comes back. It's real unstable, it's wobbly, it topples."

The erasure of people and fire from the Grand Canyon's history had created a fiction: that the dense trees, the absence of smoke, the dead fuel littering the forest floor were somehow natural. "We've created this false image of an environment that wasn't here for twenty-three thousand years," said Jason. "Now we're trying to manage it based on what the managers in the sixties, seventies, and eighties saw. It's wrong. If they had been here in 1901 when the Natives were chased out, this would have been a lot more open. There would have been probably 50 percent less fuel, not so much dead and down. There would have been humans moving around this landscape causing a lot of fires."

Jason's work in the park involved documenting the thousands of structures, lithic scatters, agricultural complexes, and villages across the Kaibab National Forest and inside the park to preserve and protect Indigenous history, which also establishes unassailable proof of Native existence and relationship to the place. "For a lot of our Native cultures," he told us, "management is existence. That's hard for a lot of non-Natives to get. But just living and moving and utilizing a landscape is part of the plan. You put humans out there, it's inevitable that there's gonna be a fire, whether someone does it on purpose or a bunch of punk kids are banging onto rocks looking at sparks. Fire follows humans; that's just what happened for twenty-three thousand years in North America."

Fire was also one of the tools Jason used to preserve this history. Prescribed fire not only re-created the human-ignited fires of the past but also ensured that the overgrown forest couldn't conceal the evidence of Native habitation, or that a catastrophic megafire didn't destroy the evidence altogether. In Jason's opinion, smoke was as integral a part of the terrain as piñon trees. "When we have these fires it gives us opportunities to look at things from a whole different perspective," said Jason. He pointed across the canyon. "We can see that the whole plateau has been terraformed. You don't see it from the highway, you don't read about it in books.... They were drawing up there. It's just everywhere. After one fire on the north side of the canyon, we went from 12 known archeology sites to 275. During another fire, we found a whole prehistoric dam. It blows my mind."

We returned to the trucks and drove further east along the South Rim, then pulled over on the side of the road. I stood and looked at the North Rim, where Steve Pyne had chased smoke for so many seasons. I texted him a photo of the canyon, its cliffs, gorges, and ravines cast in light and shadow. We followed Jason into the forest to look at an area scattered with prehistoric pottery sherds where he had prepped for a prescribed burn by removing dense fuels around a cultural site. He stopped at some long logs bleached grey. When I looked closer, I noticed that the logs were just one side of a large square laid out on the ground. It was an old Navajo structure. "This is actually a little room," he told us. "There's a corner here and here and here. I found, like, twenty pottery shards here. This is someone's house. In the one thousand years that it was occupied, it probably burned a hundred times."

Traditionally, a site like this would have been left to return to nature. Diné people, Jason explained, using the term the Navajo call themselves, avoid interacting with old things out of respect. Doing so can even hurt or make people sick if they aren't spiritually prepared. But dispossession of these lands had changed the calculus. Now Jason tried to prevent it from disappearing in a wildfire. "Without these sites, there would be no evidence of us here. If it burns, it's gone," said he. "I'm thinking in terms of seven generations in the future. If we don't speak Navajo, if we have very little of our culture left, we're going to have these sites. This might be just a collapsed pile of sticks, but they're gonna see my notes. As things change, we're always going to have to look back to these sites to see who we were, to see how we lived."

10 DRAGON EGGS AND HAMMERSTONES

THE BIGGEST CONTROLLED BURN I HAD PARTICIPATED IN SO FAR WAS A FEW HUNdred acres. In the middle of autumn, however, I packed my camping gear, fire pack, and hardhat and flew to Santa Fe, New Mexico, to join a crew burning over a thousand acres in the Carson National Forest. For the first time, I was conflicted about saying goodbye to my kids. Two days before, a case of rhinovirus had sent my seven-year-old son to the emergency room, where doctors had spent hours trying to raise his oxygen levels. He had recovered and returned to school, but now I felt overwhelmed by guilt for leaving. In the previous six months, I had traveled eighteen thousand miles, over three thousand of them by car. Now I seemed to have reached the emotional limit of my pyrowanderlust. On an empty road in the early morning, physically worn and sleep-deprived, untethered from my family, doubt began to creep into my mind. *Why was I doing this?*

My desire to be close to fire was stirring up complicated feelings. The gratification I felt from the difficulty of the work—the risk, fatigue, adrenaline—reminded me of another intense period of my reporting life, when I was living in Sri Lanka during the last year of the country's civil war. I remembered having a revelation while speeding down a highway in the country's ravaged eastern province or watching tracer bullets in the sky from my roof at night: This is exactly where I am meant to be. Now I was starting to wonder: Did I need a disaster or war to feel... fulfilled? One day I confessed this perverse thought to a friend who had worked in North Dakota's oil fields. He nodded knowingly as I described the happiness I experienced on the fireline, the compulsion I felt to return again and again. Did I know about the concept of adverse childhood experiences? he asked. He had come to believe that adults who had experienced adversity in childhood felt normal—even better—in highly stressful and dangerous situations. "It's like, 'Oh, yeah, I know how to do this,'" he said. I thought of the chronic insecurity of my adolescence, and it was like a light bulb went off. Until that moment, I had thought of wildland firefighting as being a risky occupation with the potential to traumatize those who did it. It had never occurred to me that those with trauma might also seek out fire.

I slept that night in a campground a couple of miles outside Jemez Pueblo. It was at capacity when I arrived at dusk, but the camp manager took pity on me and my nearly empty gas tank and offered me a spot underneath some cottonwoods along the bank of the river. I pitched my tent and struggled to quell my anxiety. I slept badly and was startled awake by a moon so astonishingly bright I thought someone was shining a flashlight on the tent. I packed the car just as the face of the mesa named for Guadalupe became illuminated by the blue dawn. Down the road, I pumped gas at the pueblo's station and boiled water for coffee in a Jetboil on the hood of my car. Inside, I bought a card for my kid—an image of a beagle puppy with the words "You're Awesome!" underneath—and cried as I drove away.

TO REACH THE burn unit, we passed through the small Hispano towns of Ojo Caliente, La Madera, and Vallecitos, places where for centuries people's livelihoods had been intertwined with the surrounding forest—hunting, fishing, firewood collection, community-based logging, and water use. I remembered a scene from the sociologist Jake Kosek's book *Understories: The Political Life of Forests in Northern New Mexico*. In it Kosek recounts walking with a priest named Alfredo Padillo, from the town of Truchas, not far from where we now drove, who carried a battered chainsaw on his shoulder. "You must tend to the soul as you tend to the forest. After a fire, if it is not thinned, it will grow into dense thickets, light will not be able to reach and warm the ground, and nothing will grow," Padillo said. "Sometimes pruning is the best way to foster healthy forests. Sometimes caring for God's creations means making choices."

We turned onto a dirt road and began to climb upward. To the right, the truck was bounded by steep ravines that opened onto vistas of ridges covered in ponderosa pine, Douglas fir, and white fir. On the western slope of Kiowa Mountain, we parked in a large, open *rincon* or meadow alongside a couple of dozen pickup trucks and fire engines. The grass was yellowed and dry underfoot. Everyone gathered in a big circle for morning briefing. Among the crew were hotshots, handcrews, and Forest Service employees as well as several young archeologists. They were there to mark lithic scatter sites, which contained the remnants of prehistoric stone tools, within the burn unit. The unit was so big that it was going to take two days to light it, even with the help of a helicopter. The helicopter would be dropping ping-pong balls filled with potassium permanganate powder into the unit's interior. When injected with glycol, the balls, called "dragon eggs," ignite like miniature flares and set little fires where they land that spread into one another.

The incident burn boss and his trainee stood in front of a flatbed Forest Service truck loaded with coolers and red fuel cans. A map with the name of the burn, Ensenada RX, had been taped to the truck. The burn boss picked up his radio and called in for a weather update. "Transport winds: west, eight knots," a woman reported. "Maximum ventilation rate: fair." The trainee, black hood pulled over his head, described the operation. There would be two firing teams and two holding teams, each working on the east and west side of the unit and moving southward, as well as a smoke monitor, fire effects monitor (FEMO), and the aerial ignitions team flying the helicopter. As the east side's ignitions crew began putting down a strip of black along the burn unit's perimeter, the helicopter would begin filling in the interior, littering it with dragon eggs. "If we get any spot [fires], let us know, and then hop on those things as fast as possible," the burn boss told us. "When we start dropping balls, you know, just be heads up. Monitor that air and ground pretty heavily, the rotor wash stuff. There are big snags in there, so have your heads on a swivel all day. The escape route and safety zone are back down here."

He continued, "FEMO, if you would do the weather every hour, I'd appreciate it. The bottom relative humidity on this burn is ten. If you start getting anywhere near ten, like start hitting eleven, make sure that the burn bosses and trainee acknowledge that you're getting that. Shouldn't be anywhere near there, but you never know." Then he addressed the reason we were there in the first place: "The objective, the whole reason we're out here, is to provide for the future health of the forest. We want to reduce the fuel loading to prevent catastrophic fires down the road. We're also here to restore diversity and wildlife benefit. We've got a couple [archeological] sites that can be burned through, but we don't want to trample them."

A Forest Service representative from the local ranger district stepped into the circle to speak. "I want to acknowledge that we're burning on ancestral lands," she said. "Really bringing rich, black earth, you know, making this forest what it wants to be when it needs to be." She went around the circle, thanking everyone for being present. "I'll just ask that folks be thinking of efficiencies. Any kind of wastefulness, think of the taxpayer, think of Mother Earth. I'm thinking especially of your judicious use of fuel out there. We don't need to be putting too much fuel on the ground. If it wants to go, it's gonna go. The quote that I love from a tribal member, he says, 'Fire adds richness to the land.'"

"Make sure every step you step out here is with the intention of safety first," said the burn boss. "With that, go forth, do good things. I appreciate you."

WE GATHERED OUR packs and tools. I threw mine in the back of a truck owned by a nonprofit named the Forest Stewards Guild. I was volunteering with one of the guild's crews, called the All Hands All Lands burn team, a small, mobile prescribed-fire crew whose goal was to assist federal, state, tribal, and private landowners on burns to help bring down the fire deficit.

How big was the deficit in the Southwest? By some estimates, between 1989 and 2012, the mountains of New Mexico and Arizona accrued a fire deficit of more than three million acres, and that number had only ballooned since then. Scientists had a term for this phenomenon: the "missing fire." Across the western United States, the problem of the missing fire was severe. Scientists estimated that between 1984 and 2015, 1.6 million hectares of land had burned in Washington and Oregon. But historical fire regimes indicated that without fire exclusion, 15 to 21 million hectares would have burned. During morning briefing, one of the administrators had pointed out that the unit we were burning that day had been "on the books for a while." When I asked later what "a while" meant, I found out it had taken five years of assessments and planning to finally get to the day when we could start a fire on that piece of land.

I had worked with the All Hands All Lands crew in the Zuni Mountains a few days before. We had camped on the outskirts of the town of Grants in the shadow of Mt. Taylor. In the morning, we drove into the Cibola National Forest through a piñon and juniper forest to join a local hotshot crew burning a couple of hundred acres of gently sloped ground. The fuels had been wetter than expected, and the fire was slow moving and left behind dirty, uneven black patches. It didn't matter whether I tried to lay down strips of fire or big circular dots or drew boxes or doused jackpots of downed wood; the flames never seemed to get hot enough or large enough to become a self-perpetuating blaze. By the afternoon, I had given up on firing patterns. I began to improvise, freely flinging the drip torch with my arm and wrist so the fuel scattered randomly, whirling around and walking backward, catapulting more streams of fire onto the ground in haphazard squiggly lines until I was spinning through the forest like a dervish.

Our firing boss had deemed the burn "dog shit," but the burn boss, a hotshot superintendent, actually seemed pleased. "We want to leave some logs and debris for the critters. If we clean it up too much, we'll lose the rodents and we'll lose the raptors, which defeats the purpose," he said. "I'm happy with what's left out there. Whatever we don't get is gonna melt away, and it'll be fertilizer or a hiding spot for a mouse or a family unit for a squirrel." I smiled to myself when he said this. Hotshots had reputations as snuff-snorting adrenaline junkies. But I never encountered a group of people who loved to talk longer and with as much enjoyment about fuel types, topography, trees, plants, weather, and animals. Hotshots were basically pyros who loved squirrels, I thought. When

Dragon Eggs and Hammerstones

we drove off the mountain, dusk was setting in, and as we reached the campground in Grants, the mountain we had just burned on was a black silhouette against a pink-smudged horizon. I could still make out the tendril of our fire's smoke from twenty miles away.

I WASN'T GOING to hold a drip torch in the Carson National Forest. Instead, I was training as a fire effects monitor. FEMOs, as they are called, work on controlled burns or wildfire incidents, monitoring smoke, fire behavior, effects, and weather, reporting their observations to a burn boss or supervisor to help ensure that a controlled burn is within prescription parameters. I had begun the coursework for the position a couple of months before, working my way through about thirty-five hours of a class called Intermediate Wildland Fire Behavior. Now I would work with a trainer for the day on specific assignments in my task book.

The trainer and I drove along a rutted two-track toward the east side of the unit and parked in a clearing. I checked to make sure I had my compass and a Kestrel, a small electronic wind and weather meter, around my neck and put on my pack. We hiked up a steep trail bordering the unit toward the site of the test fire. The forest understory of gambel oaks had already turned a brilliant autumnal red. The chartreuse leaves of quaking aspens fluttered in the breeze.

After a mile, we stopped in the shade of a ponderosa pine to "spin weather." The trainer handed me a small pouch called a belt weather kit that I unzipped and lay on the ground. Inside were instruments for taking weather observations, including a bottle of distilled water and a sling psychrometer, an old-school thermometer with a cloth wick on one end and a small chain on the other. I dipped the wick in the water and then stood, arm outstretched, facing the wind, and began whipping the chain in a circle so that the instrument spun around, causing the water in the wick to evaporate. The resulting number on the thermometer represented the lowest ambient air temperature at which water could evaporate. The drier the air, the more moisture that would evaporate and the lower the number would be. The more humid the air, the less moisture that would evaporate and the higher the number would be. I opened a small booklet filled with tables and began calculating the relative humidity, or RH: the amount of atmospheric moisture present relative to the amount that would be present if the air were saturated.

On a form attached to a clipboard, I wrote down the time of my observation, the GPS coordinates of the location, the aspect and slope, the wind's direction and speed, cloud and canopy cover, dead fuel moisture, relative humidity, dew point, and something called the "PIG" (probability of ignition). Then I got on the radio. Radio comms intimidated me. I hadn't mastered the language yet, whose grammar included specific protocols meant to ensure that information was crystal clear, pertinent, and always being "communicated up" the chain of command. There is a reason for this highly regimented format. Some of the worst tragedies in the fire community, including the Yarnell Hill Fire that had killed nineteen hotshots, had been exacerbated by poor radio communications. Mostly I found it hard to emulate the standard tone of radio comms, which was so laid back it sounded like surfers on valium to me. The logic seemed to be that the more radical one's disenchantment with the information being relayed, the better it would be processed and understood.

"Ensenada RX. All resources stand by for 1100 weather on tac one. Break," I said over the radio. "Temperature: fifty-three. RH of sixteen, down one. Wind direction: east. Wind speed: one to two with gusts of three. Fine dead fuel moisture: five over seven. Got a PIG of sixty over forty. How do you copy?"

"I-got-sixteen-burn-boss-and-trainee-copy-thank-you," was the reply.

We hiked toward the pale wraiths of smoke beginning to rise ahead of us. Eventually we could see ignitors dipping in and out of enormous mature ponderosa pines whose signature caramel-colored bark gave them the name "yellowbellies." The ignitors moved swiftly, navigating the rocky ground and thick gambel oaks, putting down big dots of fire. A part of me regretted not having a torch, but watching them now, I was relieved. They were already sweating with exertion. The fire behind them was burning intensely and emitting ferocious heat. A hundred feet from where we stood, we saw flames engulfing the trunks of entire trees and torching into their crowns.

"Gonna kill those guys," said the trainer pointing to some young Douglas firs. "When we were burning a few miles to the south a few years ago we were working with some guys from Tesuque Pueblo, and they didn't want the Dougs to burn. They use them for ceremony."

"This is a fire-adapted landscape though," I said. "Why don't the firs have resistance?"

"Doug firs do get fire resistant, but when they're young they get harder hit," he explained. "White firs never really get the thick bark. There is a lot of talk that the amount of white fir we see here is an artifact of suppression."

The burn boss and trainee appeared on a knob above us and hiked over, slightly out of breath. "Can you take weather every half hour instead of hour?" they asked.

"Yeah, no problem," we said.

When they left, I turned to the trainer. "We got six minutes until weather."

IT WAS AN interesting situation. To provide accurate weather observations on a burn operation, you have to find a sweet spot—close but not too close—in relationship to the fire. Too close and the fire itself would influence the readings; too far and you could be in an area with different topography, sun, or humidity from where the ignitions team was laying down fire and thus giving an unreliable forecast for expected fire behavior. Our challenge was to find a spot close enough to the ignitions team to give us accurate readings but far away enough that we had time to take readings before being overtaken by the now fast-moving flames. We started to rapidly hike downhill to outpace the ignitors.

Thus began a race lasting hours, one of us rushing to hang the kestrel, take the wind direction, spin the psychrometer, and pore through tables to make calculations, while the other peered into the forest, listening for the crack of limbs under the ignitors' feet or the whoosh of approaching flame. Once the observations were reported, we packed up and rushed again to find the next spot.

In the early afternoon, we heard the thwack of the helicopter approaching.

"All resources Ensenada RX," said the burn boss. "We're going live with the helicopter."

"AIRSHIP OVERHEAD," I wrote in the comments section of my observation form, now crowded with hundreds of numbers and notations. FEMO work was almost secretarial in its fastidious repetition. But choosing the best place to estimate cloud cover, aspect, and fuel types required careful empirical observation coupled with creativity and imagination. Was the wind truly east, or was the topography creating micro surface winds? The weather, as one firefighting handbook pointed out, was never static and always dynamic. "Outguessing Mother Nature in order to win control is an extremely difficult task," it declared. "We need to soothe her with understanding."

The smallest choices made by a FEMO had potentially outsized implications. Early on, I made the mistake of spinning weather too close to a ridge of granite rocks that had already absorbed the day's sun, resulting in an RH value so low it would have shut the entire operation down. (Five years of planning potentially wasted because of a rock, I thought.) If the burn escaped control, the form I was filling out would become a critical piece of evidence in an investigation.

"Burn boss tac one," said the firing boss over the radio. "Let me catch up with you guys to make it even."

"We got to go," said the trainer, packing up the instruments.

"They're coming at us?" I asked. The helicopter thundered over our heads.

"Yep," he said. "This is just unskilled grunt labor, right?"

"Slinging weather or FEMO work?" I asked.

"Just the whole operation," he said wryly.

In the late afternoon, we were posted up near some aspens below the ignitors. "I wonder if they got the blackline all done and are getting ready for the head fire. They might just come in and drop a bunch of balls," said the trainer. "We may need to bump to the vehicle." I called in the 1600 weather, and then we hiked to the truck and headed toward the southern end of the unit. As we drove, we passed by a meadow lined with a grove of regal yellowbellies. The trainer pointed toward a tree.

"You see that tree?"

"With the cat face?"

"That's not a cat face," he said. "It's a peeled tree, a cultural tree."

I asked him to stop and jogged over to the tree. Its bark was golden and crackled, and on its southern face was an oval wound about four feet above the ground. The wound itself was about a foot wide and a few feet long and revealed the tree's blond heartwood. Someone had cut through the bark, cambium, and sapwood and wrenched them away without killing the tree; instead, the tree had healed itself by growing a thick lip that curled inward around the cut that I now explored with my fingers.

I had heard about cultural trees before. In places like coastal British Columbia, First Nations people peeled trees for carvings, canoes, and weaving. In Australia, carved trees acted as navigation aids or marked spiritually significant places like birthing sites or placenta burials. Cultural trees in the Southwest were relatively rare and something of a mystery. Archeologists and dendrochronologists have hypothesized that the nutritious cambium layer of the pines might have been harvested during times of famine or war or forced displacement. Analysis of their tree rings indicates they were used throughout the eighteenth and nineteenth centuries by Apache, Ute, Navajo, and Zuni people. There was also speculation, based on stories passed down by Indigenous women, that the peeled cambium was used as an abortifacient. As I looked at the scar, it occurred to me that famine and the need to end a pregnancy and to leave behind a mark of one's existence in this forest might have collided all at once.

Equidistant between the east and west firing teams, we spun weather between two giant plumes of smoke. I radioed in an observed increase in the wind to five to seven miles per hour with gusts of nine. We drove further west along the perimeter. The winds quieted and the relative humidity rose. We took our last weather observation at 1800, and the burn boss shut down firing ops for the day. I began to think about where I was going to put my tent that night. But as we sat in the cab, I became aware of a flurry of radio traffic.

"Did someone just say there's a spot?" I asked.

"Yeah," said the trainer. "A big one."

We drove as fast as the dirt track would allow, then parked in a turnout and gathered our packs and tools. The daylight was waning into a purple dusk, and I pulled out my goggles and headlamp and stretched them over my hardhat. The spot fire had ignited across the east side's perimeter on a steep slope, and we hiked at a clip and then turned sharply into the forest, using the clanking of tools hitting rocks and the whining of chainsaws as a guide. Eventually I saw the blue beams of headlamps and the glow of an active fire burning on the ground and in heavy downed logs. I got into line behind the other firefighters and began to dig. The soil was thin and matted with grass; every strike hit the underlying rock with a clang and made my tool vibrate. When the hoses arrived and the spot was finally contained, I hiked back down to the dirt track and waited till everyone had passed me by. I turned off my headlamp. In the blackness of night, the burning forest became mesmerizing. The floor gently flickered with thousands of points of white light like a galaxy of stars had draped across the earth. The sound of crackling wood surrounded me in an ambient ether, and I walked back to the truck as slowly as I could.

THE FOREST SERVICE liked to point out that it oversaw the country's largest prescribed-fire program. The agency burned almost a million and a half acres annually. Of the roughly forty-five hundred fires it ignited each year, fewer than 1 percent of them escaped control. The ones that did, however, were often emblazoned in collective memory and cited for decades. People didn't forget, let alone forgive, when the government intentionally lit fires that ended up burning down homes. This was especially true in New Mexico, where in 2000 the Cerro Grande Fire, a prescribed burn in Bandelier National Monument, escaped control and became one of the largest recorded wildfires in the state.

The story of how the fire grew out of control illustrated how quickly things can go wrong on a prescribed burn. The National Park Service set out to burn a thousand acres in the headwaters of Frijoles Creek, which lay in a forest of mixed conifers and ponderosa pine. The goal was much like it was for the burn I worked on in the Carson: to reduce fuel loading and restore fire as a keystone natural process. The crew started their test fire in the evening. Two firing teams began at a northerly point, one working east and the other west. Within a couple of hours, the fire showed signs of burning more aggressively than anticipated and slopping over the perimeters of the unit. By the next morning, a Type 1 crew, air tanker, and helicopter were requested to hold the fire, but by the afternoon a spot fire was detected a quarter mile away. Over the coming days, the fire spread across tens of thousands of acres. It would take a month before it was fully contained, and during that time it ran through Los Alamos, the site of the country's largest nuclear research laboratory. Ultimately, the Cerro Grande Fire burned forty-eight thousand acres.

From an ecological perspective, the Cerro Grande Fire was neither wholly bad nor good. As one scholar put it, "The fire burned large acres at low severity/intensity, large acres of high severity/intensity, and large acres of a mixture of interspersed severity/intensity levels. The ecological implications of these fire behaviors are equally diverse." Socially, the Cerro Grande was a disaster. It destroyed 235 homes and cost an estimated billion dollars in damages. Although Los Alamos National Laboratory claimed that no nuclear material was affected, the fire burned through areas known or suspected to be contaminated with radionuclides. According to Jake Kosek, many local people believed there was a direct connection between the Cerro Grande Fire and the mass die-off of piñon trees in the years that followed. While ecologists attributed the death of millions of trees to drought and warming temperatures, locals were convinced that the piñon forest had been blanketed by radioactive ash. The fire also shut down prescribed-burn operations across the country for months. "The Cerro Grande Fire in New Mexico has become one of the worst examples of 'prescribed' fire gone wrong," declared the *Denver Post*.

I thought about the Cerro Grande Fire as I drove back through the tiny, sleeping villages for a second day of burning in the Carson National Forest. What if we hadn't caught the spot the night before? What if it had become another Cerro Grande? How much would the fire deficit grow in the time it took to restore the public's faith in the necessity of good fire?

It was always easy to defer the risk of prescribed burning to some future time. Yet not burning would only increase the fire deficit and create new risks in the form of megadroughts and megafires that put landscapes and communities in peril. The cost of containing a megafire often dwarfed the costs of an escaped burn. The Cerro Grande Fire had cost $56 million (adjusted for inflation) of taxpayer money to suppress; twenty years later, the Dixie Fire had cost $637 million. Some scientists felt the time had come to mandate prescribed fire in order to tip the risk calculus in favor of controlled burns. "When a wildfire starts, the piggy bank opens. No one is managing where the smoke goes; no one questions when fire jumps out of containment," I heard one burn boss argue. "The rules of engagement are vastly different [between wildfire and prescribed fire], but they shouldn't be."

Fire seemed to demand a radical departure from the norm in our perception of risk—to require us to think in one-hundred-, five-hundred-, and thousand-year timescales. Could we extend our care into the future for a world that didn't yet exist? If so, the right question to ask was not "What if the fire escapes?" It was "What will happen if we don't burn at all?"

We gathered in the same meadow for morning briefing, though now the air was rich and pungent with yellow haze. The crew had burned 730 acres the day before, and there were still a few hundred left. "Watch your footing and snags," said the burn boss trainee. "Stuff's been burning through the night, so some of these aspens and dead logs are fire-weakened, could come down any time." I wasn't training as a FEMO that day, so I set out with my pack and tool, gridding in a staggered pattern through the forest with other firefighters in a search for any more spot fires that might have smoldered through the night. When we finished, I made my way over to the west side of the unit to join the holding crew, following behind the guys with drip torches as they began working their way down the steep perimeter of the unit and making sure their fire didn't blow across the line.

It was drier than the previous day, and the fire quickly intensified. As I watched firs torching, I was reminded of an entry in the diary of Meriwether Lewis, who wrote in 1806 in what would later become Idaho that their Native American hosts had spent the evening entertaining them by "seting the fir trees on fire. they have a great number of dry lims near their bodies which when set on fire created a very suddon and immence blaze from botton to top of those tall trees. they are a beatifull object in this situation at night. this exhibition reminded me of a display of fireworks."

The western perimeter of the unit was a curvy squiggle, creating opportunities for tricky wind shifts and for backing fires to become flanking fires and upslope head fires. The ignitors worked slowly, laying down a bit of fire, moving out of the way, watching, trying to temper the fire's behavior and speed. The ground they left behind was littered with thousands of cones turned to white ash. When the ignitions crew rounded one particularly sharp corner, the firing boss reported over the radio that they were observing a fire whirl and waiting for it to dissipate. The smoke plume became enormous, an undulating orb of brown and blue vapors that shape-shifted and rose into the sky.

Mostly, I stood with my back to the fire, downy ash falling around me like snow. I scoured the ground for the littlest wisps of smoke indicating that an ember had blown across the line into the green. I inhabited a weirdly bifurcated mental space: while part of my mind was totally focused on a single task, the rest of it seemed liberated to wander through strange and dreamy digressions. The hours passed. As the sun began to sink slightly in the sky, I saw a single aspen growing amidst the pines. Its leaves were a shockingly alchemical yellow, and they trembled and vibrated in the breeze. The aspen's crown seemed to be emanating all the light it had ever absorbed outward into the world. As the sun sank lower in the sky, it became a luminous disco ball shooting its beams through the scattered smoke particles.

I had been told to hold up at the southwest corner of the unit. When I reached it, I cooked noodles in a Jetboil on the tailgate of a pickup truck and snapped some branches to use as a pair of chopsticks. I sat on the ground and began to eat.

I don't know how long it took me to realize the rocks surrounding me were unusual. It happened slowly, as though my unconscious mind were discerning that there was something interesting about how they were strewn about, that they were not entirely random in their size and assortment. When I finally got up to figure it out, I saw a tree nearby wrapped in three pieces of white tape. I had been sitting in one of the archeological sites. I walked around the rocks slowly, stopping for long moments. Could I recognize signs of the Paleolithic people who had once gathered here to knap spear points, scrapers, and hand axes? I looked down at a lump of grey chert with patches of mint-green lichen growing on it. It had a slightly inorganic rectangular shape. I sat down on the ground and looked closer. On its upper right corner, the rock had a gentle convex plane where a piece of it had broken—or been broken—off. I turned the chert around, trying to divine the purpose of its odd angles. Then I flipped it over. Its underbelly was round at the bottom and came to a point. There was a gentle divot there. I picked it up and felt the rock's shape slip into the curve of my palm and my thumb fit into the divot like a key into a lock. Was this a hammerstone? Whoever had held it before me had a hand exactly the same size as mine.

Clovis people, named for an archeological discovery at a town just a few hours away from where I sat, had hunted mastodons and sloths across this land as far back as thirteen thousand years ago. The site I explored might have been at least a few thousand years old or even older—I wasn't really sure. How much had the forest changed since then? Did the people who sat here striking these stones marvel at the aspens too? All around me was proof of their existence, evidence that humans were as much a part of the landscape's past as they were a part of its future.

11 BEAUTIFUL AND RIGHT

IN THE FIRE COMMUNITY, THE SOUTH HAS A REPUTATION AS THE PLACE ONE GOES to burn big and often. No other region in America sets fire to more acres each year. Across the entire West, some 2.5 million acres are burned annually; in the South it is over 6 million. Two million of those acres are in Florida alone. As a testament to the state's flammability, I once heard someone claim they had watched water burn in the Everglades, a seemingly miraculous phenomena of flames skipping across the swampy terrain. The historian Steve Pyne wrote, "In Florida fire season is plural, and it is most often a verb. Something can always burn; something almost always does." The South was where fire traditions had been stubbornly upheld, and it was those folk arts and generational knowledge that people drew on when they sought to reintroduce fire to the rest of the country. The South had states where a landowner's right to burn was enshrined in law. Fire lighters are protected from civil liability unless they are found in court to be "grossly negligent." "There's literally bizarro levels of fire here," someone told me.

The South was the only place I heard of where "fire festivals" are held, annual events that bring people together to celebrate fire and prescribed burning. Families play with fire equipment, go on hayrides, and watch live fire demonstrations. I got a taste of the region's radically different fire mores after signing up for a mentorship through the International Association for Wildland Fire. I was paired with a third-generation woods burner and wildfire biologist with the South Carolina Department of Natural Resources by the name of Johnny Stowe. My correspondence with Johnny had only confirmed my sense that when it came to fire, the South was another universe. Johnny liked to say we were all traveling on what he called a "pyropathway," a journey to understanding sacred fire. He had grown up burning his family's longleaf pine forests in northwest Georgia and managed heritage preserves in Alabama and South Carolina. He liked to sign off his emails with "STF," which stood for "sharing the flame." He called fire the "ecological imperative" and intentionally mimicked the lightning strikes that shaped his homeland's longleaf pine ecosystem. "I hope to be lovingly lighting fires and basking in their beauty on my favorite preserves on Monday and Tuesday," he would write, "if it doesn't rain before then and I get some wind and lower humidity." Other days he told me he was "wide-open as a peanut hull burning or prepping land and equipment for burning." Johnny liked to quote William Faulkner as much as Aldo Leopold, and it was his belief that natural resource professionals in the South needed to reconnect their profession to the culture, heritage, and character of the places where they worked. "We allowed the ancient multicultural tool of woods-burning to be usurped by a cartoon bear and other carpetbaggers," he argued. On his own family's land, he sought to keep alive the rituals of his ancestors: hunting, timber harvesting, firewood cutting, restoring grasslands, and burning. "Some of my trees are harvested and some are hugged," he would say.

When the burn season starts in the South, wildland firefighters from all over the country looking to make extra money in their off-season head there in a pyromigration. "It's like a vacation for them," a hotshot told me. It became one of my fervent desires to go south in order to experience its "peculiar fire heritage," as Steve Pyne put it, to be in a place where fire-lighting approached something more akin to art than science. I started asking around: Was there a crew I could join for a week? At the end of February, I got word that there was room for me in Wilmington, North Carolina, where a seasonal crew was burning in a variety of ecosystems: pond pine forests, evergreen shrub bogs known as pocosins, hardwood bottom swamps, and longleaf pine savannas. I packed up and left New York before dawn one Sunday for the ten-hour drive.

THERE WAS A time when the longleaf pine forest started in Virginia, stretched all the way to Florida, and spread east to Texas. Across ninety million acres, it was a woodland savannah drenched in sun. The understory of wiregrass abounded with carnivorous plants, orchids, sedges, and wildflowers. In any thirty-foot square of longleaf pine savannah, there might be eighty unique species of plants growing. Without fire, the longleaf stops regenerating; the grasses and plants are crowded out by shrubs. Lightning was as critical to the forest as soil, temperature, rainfall, and wind. In 1902, the naturalist Roland Harper took a photograph of a young child in a longleaf forest in Alabama. The child is staring up at a giant tree while touching its trunk with both his hands. Surrounding him are hundreds more trees equally majestic stretching far into the distance. The earth around the boy's feet is grassy and covered in a play of light and shadow. It is a forest that invites you to run with abandon, to play a game of hide and seek.

The wood of a mature longleaf pine tree is dense with tight growth rings and a resin-soaked core known as heartwood. Settlers used the trees to make pitch, tar, turpentine, and rosin, and in the nineteenth century, trees were cut and floated down rivers to market. Eventually, railroad lines were built directly into the forest, and the trees were loaded onto train cars. By the turn of the nineteenth century, millions of acres of virgin forest were being clear-cut. Ten years after Harper took his photograph, he wrote, "These trees were doubtless cut long ago, and it may never be possible to take such a picture in Alabama again." No one considered the longleaf pine forest a renewable resource; the approach was "cut out and get out." The loss of longleaf forest in the twentieth century was said to be even greater in scale or impact than the disappearance of the American chestnut.

Someone who caught a glimpse of this primeval forest before it disappeared was the ecologist Herbert Stoddard, who moved to east-central Florida as a child in 1893. The timber industry hadn't arrived yet, and it was common at that time, as Stoddard put it, for "wimmen folks" and children to set the annual fires around their homes and in the woods. He described the work as

> hard, dirty though interesting.... Great care had to be taken to avoid damaging the rail fences that encircled each corn, cane, sweet potato, and cowpea patch, and the ever-present cow pens. It was hard work to split rails from even the straightest-grained virgin longleaf, so the "snake fences" were well worth protecting from fire; the pitchy rails burned fiercely, and the blaze could not be easily extinguished.

One day, the five-year-old Stoddard decided he wanted to light fires, too, and torched the grass near his house. But it was a dry and windy day; the fire he started burned for three days and nearly consumed his family's home. "The southern people that lived nearby took it in stride as a matter of course," he wrote.

Stoddard worked with the local cattlemen, whom he seemed to be in awe of. Frequently maligned for setting fires, they rode bareback and barefoot through forests that were kept open and spacious and abounded with huckleberries and blueberries, semiwild hogs and bears. At night, cattlemen set out together to harvest wild honey from marked "bee tree[s]." "The frequently burned-over woods were easily traversed on such expeditions, and they well knew that the burning increased the growth of Partridge Pea and the flowering of the Saw Palmetto and other honey-producing plants," said Stoddard. He believed that "the early pioneer cattlemen of central Florida had neither horns nor spiked tails. Since nobody had come to their defense in the last half century, I felt an obligation to do so."

Stoddard was a high school dropout and never attended college. He worked as a taxidermist and ornithologist before returning to Florida in 1927 to undertake a study of bobwhite quail in the Red Hills around Tallahassee. By that time, many of the virgin stands of longleaf pine in the region had been cut down, and young longleaf pine struggled to grow back in the "rough" because no one was burning it anymore. During the Great Depression, the Forest Service bought up defunct timberland and sent the Civilian Conservation Corps out to reseed it, but without reintroducing a regime of intentional burning. During droughts in the 1930s and 1950s, massive wildfires broke out. "I have always felt that the so-mistaken anti-fire propagandists were largely responsible for these great conflagrations," Stoddard later said. In 1931, Stoddard published what seemed like an unassuming book called *The Bobwhite Quail: Its Habits, Preservation and Increase*. In fact, the book's findings were radical. The book not only helped define the profession that would come to be known as wildlife management; it advocated for the idea of ecological diversity and presented evidence of the critical need for fire in the savannah-like longleaf pine forests. Officials at the Forest

Service hated the book so much they made Stoddard rewrite its fire chapter half a dozen times before they would publish it. (The same year it was published, Congress passed the Clarke-McNary Act, which withheld federal funding from states if they engaged in or allowed fire-lighting.)

Stoddard wasn't the only person who recognized the need for fire. Another was Henry Beadel, an amateur naturalist who, like Stoddard, had first roamed Florida's longleaf pine forests as a child. In 1894, Beadel's uncle had bought a former plantation near the shores of Lake Iamonia, a "happy" place, according to Beadel, of fields, pine groves, and hammocks, live oaks, gums, and maples. Obsessed with hunting quail, Beadel and his brother set out every afternoon with Charlie, the plantation's Black land manager, to shoot birds. One day in late February, the boys witnessed a horrifying sight: a fire burning through the trees. It looked like the whole countryside was on fire, but Charlie explained that the fires had been lit on purpose. People had been doing it every spring for as long as even his great-grandpa could remember.

Beadel took over his uncle's land in 1919, and the old burning routine was still going on. The tenant farmers raked around the buildings and fences, and on the last day of quail hunting, they would start burning. "That night," said Beadel, "on every hand, lines of flames crept or raced across fields, flickered through pine woods, here and there flaring high over the heavier clumps of weeds, accompanied by cracklings of brush, bangs like pistol shots, and clouds of eye- and nose-stinging smoke." But Beadel wondered if the annual burning was an artifact of bygone days and really necessary. He decided to stop the tradition. After three years in which the quail population dropped, he was convinced that the old-timers had understood the necessity of the flame's effects better than him. "The fact that this region developed and has maintained a condition of ecological stability through many, many years of uncontrolled burning is incontrovertible proof, to me," said Beadel later in life, "that fire is not the timber-vegetation-game-destroying demon as it has been so often pictured. On the contrary, it is an essential factor in maintaining that stability."

Stoddard's and Beadel's stories became intertwined in 1958 when they launched a research institute at the old plantation. Called Tall Timbers, its purpose was to study and practice the management of longleaf pine forest, principally with the use of fire. It began hosting annual conferences that brought together the naturalists, scientists, botanists, and land managers who would foment a fire revolution.

Setting fires in the woods was still publicly deplored. Indeed, it would be hard to overstate how maligned people were for their belief in the goodness of fire. Native Americans and Black tenant farmers whose forebears had often been enslaved on the same plantation land they now burned, backwoodsmen, bareback-riding cattlemen, hill folk, and women were shunned, castigated, and criticized. Their tenacity to set fires was the only reason the practice survived.

The first person to ever publicly testify to the need to burn longleaf pine was a woman. In 1888, Ellen Call Long, a Florida silk cultivator and preservationist, stood before the American Forestry Congress. "There is sound reason for believing that the annual burning of the wooded regions of the South is the prime cause and preserver of the grand forests of *P. palustris* to be found there," she told them. Her profire view was so unpopular that Long said she had been almost afraid to read her paper aloud. Her testimony was forgotten until the ecologist Roland Harper finally gave her credit at the first Tall Timbers conference. Long "must have been a very brilliant woman, sound[ing] one of the first discordant notes in the chorus of denunciation of fire," he said. "It was so heretical that little attention seems to have been paid to it at the time."

Ed Komarek, another cofounder of Tall Timbers, liked to tell a story of a high-ranking Forest Service official who visited Florida in the 1930s. While driving around with the official, Herbert Stoddard spotted one of his friends out in her yard, a Black woman by the name of Miss Sally. They stopped to talk with her, and the official said, "Somebody's been burning around your house here."

"Yes, sir," she responded.

"Did the plantation do that?"

"No sir," she said. "I did it. They let me do that, so I burned a couple of days ago."

"Well, look at it," said the official in consternation. "You burned everything out there. Just look at it. That land that you burned, it's just as black as you are."

"Yes sir," she said, "and in about three weeks it'll be just as green as you are!"

OUTSIDE OF WILMINGTON, the air became thick and grey, and then the rain started to fall. The downpour lasted for hours, and I worried that the moisture might delay the crew's burning. When I met the crew the next morning, though, everyone was excited. The wet weather was a boon for them. It had been so dry that spring; low relative humidity and gusty winds kept shutting down their burns. Now, thanks to the rain and sodden ground, we would be able to light fires all week. We gathered each morning in a giant warehouse on the outskirts of the city, where the crew's engines, UTVs, ATVs, chainsaws, drip torches, fuel, and tools were stored. The crew was split into several modules of four people. My module leader was Drew Ziglar, a former hotshot with sandy-brown hair and a drawl. When he wasn't burning or taking people out fishing on his charter boat, Drew was surfing big waves in Indonesia, Hawaii, or North Carolina's Outer Banks. He joked that he was a "hippie redneck."

On our first day we hooked a trailer with a UTV to the back of the crew truck and headed south along the Cape Fear River. Our first burn unit was a piece of land under a conservation easement on the border of an eighteenth-century plantation. Drew drove, one arm slung on the wheel, and told me how a hedge fund manager who was a direct descendent of the plantation's original owner had bought it recently. There was a local rumor that as part of the renovation, they had paid a million dollars to move an ancient live oak because they wanted it in a different place. "When you want to sit in your rocker with your sweet tea," said Drew, "I guess it's worth it."

We turned onto a dirt road bifurcating dense woodland and parked. It was already warm, and the crew took off their shirts and put their yellows on over their bare skin, leaving them untucked to get a breeze. We started organizing equipment. All the UTVs had names, given to them by the smokejumpers who used to come to Wilmington in their off-season. I noticed one of them was emblazoned with the words "Pony Boy."

"Who's Pony Boy?" I asked.

"I don't even want to know," said Drew, shaking his head.

The crew grabbed leaf blowers from the back of the truck. Every region has a quintessential fire tool. In the prairie, it is the flapper. In the West, it is the Pulaski. Below the Mason-Dixon, it is the leaf blower, a surprisingly effective device for clearing needles, leaves, and detritus off a line. The blast of air can even direct or blow out flames. Drew grabbed his blower's pull cord, and it roared to life. "Welcome to the dirty South!" he yelled and began sweeping a roadside ditch.

I walked a ways into the woods to try to figure out what we were actually burning. For outsiders, North Carolina's landscape is bewildering. It contains hundreds of natural communities made up of unique assemblages of particular plant, soil, moisture, topographical, and fire interactions. The ecologist Mike Schafale has compiled an inventory of every natural community type in the state; it contains thirty different categories with hundreds of subtypes. There are sand-myrtle heath balds and catawba rhododendron heath balds; northern prairie barren xeric hardpan forests and acidic xeric hardpan forests. Schafale has found fifteen different variations of dry longleaf pine communities and thirty subtypes of coastal plain floodplains.

The woods I walked through had pond pines and loblolly pines, tall and straight as arrows with punky green crowns. Underneath the pines was a crowded understory of shrubs, many taller than my head. I was in pocosin, a kind of evergreen bog that has its own special assemblage of shrubs: fetterbush, titi, gallberry holly, honeycup, red bay, and loblolly bay. Or, as one of the crew put it, "Pocosin just means really thick shit you don't want to go through." Pocosin was some of the wildest habitat in eastern North Carolina. According to Schafale, "pocosin crashing" for a day could make you feel like you had climbed a mountain or wrestled a bear. Pocosin typically had a soggy ground created by the black muck of organic matter that accumulates over thousands of years and creates a carbon sink. It is also the primary habitat for carnivorous flora: pitcher plants, sundews, and bladderworts. Until the 1960s, around two million acres of North Carolina was covered in pocosins. Today, it's around seven hundred thousand acres; most of it was drained for timber plantations or agriculture or transformed as a result of fire exclusion. Typically, pocosin habitat would have burned every five to thirty years, a regime that creates a complex mosaic of closed- and open-stand structure with incredible species diversity. Without fire, pocosins lose the nutrient release produced by combustion and become overrun with woody shrubs. They also become tinderboxes.

This was my main thought now as I stared at the crowded understory. Some of the "shrubs" were ten feet high. Pocosin is notoriously difficult to treat with prescribed fire. "The behavior of fire in pocosins is particularly dramatic for its intensity, its unpredictable behavior, and its paradoxical effects," according to Schafale. "But in the spring, when the new leaves look the freshest and wettest, the vegetation is volatile and much more likely to burn. And during droughts, all bets are off." At the start of the year, 99 percent of North Carolina had been abnormally dry, with over 40 percent of the state experiencing severe drought. During a previous severe drought, in 2008, a lightning strike in pocosin started a peat fire that smoldered for three months through forty thousand acres. How exactly is this going to work? I wondered.

The crew gathered round for briefing. The plan, the burn boss trainee explained, was to burn the unit as fast possible. "It's an east today and a west tonight, and directly to the east of us is Highway 87 and a middle school and a high school," explained the trainee. "We want to let this thing cool down as long as possible before that wind shift." Then we covered the objectives of the burn.

"One hundred to 80 percent killing shrub species of two inches or less," someone offered.

"Burn 80 to 100 percent of ground cover," said someone else.

"Our LCES"—lookout, communication, escape route, and safety zone—"is that we're all our own lookouts," said the burn boss trainee. "If you can get through the shrubs to get to the drain, that's a great spot. But we should be bringing nice black with us today."

"We're just going to pull strips, stay tight together, just trying to build the heat and make sure we can get things hot and burning as good as possible," said Drew. "If the burn's good, we'll adjust and widen things out. We're trying to light this off at once."

We lit the test fire in the western corner of the unit, and within a few seconds the flames backing through the shrubs were six feet high. I could never get over how odd it was to be just steps away from flames that could kill you, to see the border between life and death physically manifest. Standing at the edge of a fire was like standing on the edge of a waterfall: you couldn't help but contemplate what it would be like to jump. We set out with our drip torches through the unit, bushwhacking and then stumbling into surprisingly open spaces. It was pretty in an odd, otherworldly way. Around my feet, airy balls of mint-green deer moss and carnivorous pitcher plants littered the ground like terrestrial coral reefs. The smoke smelled of caramel and hog fat and citrus. I watched as the flames tussled the green heads of young longleaf pines. We wrapped the entire unit and stood watching the smoke rise; it took just forty minutes to burn thirty-eight acres.

"The fact that we're between two highways and a middle school and I didn't get a single call and the fire department didn't show up is amazing," said the burn boss trainee afterward.

"Talk about a fire culture," agreed the burn boss.

In the days to follow we set out each morning to burn an array of landscapes. One day it was an old sod farm that had been put under a conservation easement and was going to be reforested with longleaf pine. "In, like, fifty years, it's going to be sweet," said one of the crew. The next day, we burned a seventy-acre preserve of longleaf pine, many of the trees banded white to mark red-cockaded woodpecker nesting sites. I dragged torch in slurping mud up to my knees and choked on smoke putting out slopovers. These moved faster than anticipated, and sometimes I found myself running to extinguish them without having grabbed a tool or even gloves, just stamping the flames with my boots in a demented jig. One afternoon, I was kneeling on the ground to fill a drip torch when the translucent gas caught the sunlight and shimmered like mermaid scales. I accidentally inhaled the fumes, which seemed to pierce my brain. I swallowed ibuprofen and handed the tablets out to others like they were Tic Tacs.

On the way to a burn one morning, we passed a church whose sign said, "Pray to the Lord for the healing of the land." I listened as the burn boss described all the places he had traveled to light fires. One of his favorites was Cape Town, South Africa, where he had set fires on the slopes of Table Mountain. "But," he said, "I always wanted to burn in Australia. I heard they burn in flip flops."

"I heard they just burn it off into the ocean," chimed in one of the crew. "Hundreds of thousands of acres."

This kind of talk was familiar. Fire lighters seemed to carry with them a dream of a place where fire was stripped of safety regulations, government qualifications, diesel fuel, and engines. In this pyro-Eden, fire was so benign that igniting one was like watering a garden.

THE ANNUAL CONFERENCES hosted by Tall Timbers became a meeting place for pyroevangelists. At the second one, in 1963, the anthropologist Omer Stewart spoke. He had been personally invited by Ed Komarek, who considered Stewart a "fire maverick." The anthropologist presented a paper entitled "Barriers to Understanding the Influence of Use of Fire by Aborigines on Vegetation." The main problem, Stewart told the audience, was the prejudice of scientists and ecologists who rejected the folklore of the "peasants," "country-folk," and "black Negroes" by assuming they could contribute nothing to modern scientific inquiry. "Not only scientists, but all whites of European ancestry have always found it difficult to take the Indians seriously enough to learn from them," said Stewart. "The relationship between Indians and whites started with the assumption that the Indians were only part of the natural environment." Stewart argued that scientists had made the grave mistake of assuming that their observations and studies were of "Nature," when in fact they were of places where people and nature had interacted with one another for eons.

Stewart was preaching to the choir. Tall Timbers brought together people who revolted against the idea of roping off places for conservation. As the ecologist Leon Neel recounted in his memoir, "The late 1950s and 1960s was a period of growing support for the preservation of wilderness, and Americans of all stripes were increasingly committed to preserving pristine landscapes from human exploitation and manipulation. We were actively maintaining a system that would have degraded without our interventions."

Neel was a young forestry student when he met Herbert Stoddard and became his acolyte. He had grown up in the woods of Georgia. As a young kid, he once walked through a briar patch and thought to himself, "Shoot! This thing needs burning off!" Neel considered it his first prescribed burn even though the fire he started had been far from controlled. Stoddard taught Neel that "land management was an art based in science." It combined basic knowledge, careful experimentation, adjustment, and, as Neel put it, "what you come to know as beautiful and right." Over three decades, their approach to forestry combined ecological management with silviculture. Their goal was to mimic natural-disturbance regimes such as blowdowns and lightning fires through timber harvests and prescribed burns in order to create variegated-aged stands of trees with healthy grass understories. Whereas conventional silviculture practices were to cut down mature longleaf pines once the trees reached a certain age and stopped adding volume, Stoddard and Neel encouraged old-growth trees even if they weren't profitable; under the right conditions, some trees might reach four hundred years old. Indeed, Neel argued that you could create a healthy forest and still make money from it: he estimated that between 1945 and 1995 he harvested more than fifty-six million board feet from lands under his management while *increasing* the standing timber by forty-eight million board feet.

In his personal history *The Art of Managing Longleaf*, Neel explained that their management approach grew out of an aesthetic appreciation of the open woods. *Why did it look the way it did?* They recognized that Indigenous burning had profoundly influenced the forest's composition and elements over the course of its existence. "The forest as we know it thus took shape in concert with human action, and it is difficult to say whether natural or anthropogenic fire had more to do with the system's development," wrote Neel. To them there was little difference between aesthetic and ecological values. Aesthetics informed the art of management. "In fact, to a great extent it *is* the art," wrote Neel. But it was rooted in an acknowledgment that no ecosystem is fixed, and none should ever be preserved or managed as though it were. "Everything out there is alive and growing," wrote Neel. "Living things are being born and things die; it is just not static. Conservation must be exactly the same way."

Fire was the Stoddard-Neel approach's most powerful and complex tool. The two men's aim was to create a forest whose structure would support repeated burning over a long timescale. To those that paid attention, fire offered revelations. It helped you understand the past and make decisions about the future. Fire "tends to the complexities of the longleaf system," wrote Neel. "In a sense, when fire is applied and managed well, it does some of your thinking for you."

At Tall Timber conferences, fire-lighting was frequently spoken of in terms of art. Henry Beadel referred to practices like setting fires at night as the "art of advanced burning." He swore that some people could "make fires travel straight against the wind." Komarek spoke of the "folk-wisdom" of pioneer peoples and Native Americans who, generation after generation, developed an understanding of the use of fire that was akin to art. "Man has used both the science and the art," he said. "The two are so closely entwined that those of us who use fire must be both scientist and then artist, or artist and then scientist."

Komarek spoke those words at the seventh Tall Timbers conference in 1967. That year, the conference was held near Hoberg, California, close to where Harold Biswell at the University of California, Berkeley, had been conducting his controversial experiments burning in ponderosa pine forest. One of Biswell's graduate students, Jim Agee, was in attendance. The experience was so extraordinary, Agee still remembered it fifty years later. When Komarek took to the stage, he began to passionately describe the qualities of fire. "It can travel slowly, quietly, and be as gentle as the whisper of a breeze; or it can travel with tremendous speed, literally roar, and be as destructive as the greatest of storms; and it can change from one to the other and even back again, time and time again with near lightning speed." As Komarek extolled the virtues of fire, he began shouting and waving his arms. He described the evolutionary history of fire on earth, man's desire to play with fire, and its crucial role in ecology: "In nature, fire is a great regenerative force, one might even say rejuvenative force, without which plant and animal succession, in the absence of climatic upheaval or physiographic cataclysm (or at least of great climatic or physiographic change), would be retarded so that old, senescent, and decadent communities would cover the earth."

As Komarek spoke, someone else in the audience rose to second his words. Then his wife, Betty Komarek, a botanist and prescribed-burn manager, stood up and did the same, followed by someone else. "It was a revival meeting in all its glory," Agee told me. He was so stunned by the displays of emotion that he wondered if he had chosen the right profession. "I was interested in fire science, not fire religion," he said. "I soon concluded that fire people from the South simply had a more bombastic way of expressing their thoughts about fire, in contrast to the more reserved approach of the western fire folks." At the end of the conference, everyone went up into the mountains to watch Biswell put on a controlled-fire demonstration. According to Neel, "That fire stopped right at the top of the ridge, just like he wanted. He was an expert at using natural conditions to control fire."

Supposedly, Herbert Stoddard loved to ignite fires into old age. At seventy-seven, he still carried a drip torch, walking along a trail and setting a head fire. He didn't wear Nomex or a helmet or carry a radio. He wore old pants, and it was Leon Neel's job to watch for when the fraying cuffs would inevitably catch fire so he could put it out. Stoddard thought that fire management was all about calculated risks. "I don't think we can possibly ever use fire on a large scale and get the results we ought to get unless we do take those calculated risks," he once said. Timidity, he thought, was the enemy of fire. "[Some men] are just not temperamentally suited to it," he said. "They are too timid, and that's one thing I learned long ago that there's no use trying to make an expert man in the use of fire who has a timidity in his whole makeup."

TOWARD THE END of the week, Drew told me our module was going to go to Pamlico County to help out a state Wildlife Resources Commission crew. It was a three-hour drive, and we left early, hugging the coast and then skimming the Croatan National Forest toward the Pamlico Sound. Beside me, a guy read a Murakami novel, and I listened to the idle chatter between Drew and a young wildland firefighter who was about to start his rookie season on a hotshot crew. The summer before, the young guy's handcrew had been assigned to the KNP Complex, and now he described seeing the bark of burning giant sequoias slough off the trees like lava. The conversation drifted to hotshotting. Drew was still torn over whether he should go back to it. He reminisced about his last day with the crew. They had done a massive prescribed-burn operation, and he had shot incendiaries from the edge of a ridge into the sun, smoke and light bouncing off each other. "I kept thinking, 'Yeah, I'm going to do this forever,'" he said. "Then, when it was over, I was like, 'Nope, I'm done.'"

We turned onto a long access road of mowed grass running between the woods, and parked next to a group of trucks. I met Travis, the forest manager and crew boss.

"A writer?" said Travis, when I introduced myself. "Well, I'm a redneck." Everyone laughed.

We installed a pump in a nearby pond to fill up the engines and then stood around for briefing. We had eight people to burn ninety acres of longleaf pine that had become choked with slash pine and brush. "It's going to be *a lot* more volatile than [just] longleaf," said Travis. Two sides of the burn unit were bounded by water: one with a brackish estuary that led to the Pamlico River, and the other by a marsh with a wide creek running through it. "Predominantly, we got a west wind blowing across," said Travis.

> Basically we're going to get this lit all the way to the water, wet line that little bit. Probably what we'll do is come back up here and start dotting across that way; we got closed canopy here in the midstory, a lot of regeneration. I want to try to save it so we can generate some money off a timber sale. Space it out, let these fires come together so we don't build up so much heat underneath to kill it. The biggest spotting potential is up there at the gladed part. Hopefully we can attack anything we got. We'll have to finish off lighting the marsh with the Marsh Master.

The Marsh Master. I had heard it mentioned with excitement on the drive. Now I turned and eyed a strange amphibious creature stationed in the grass. It looked like someone had smashed a boat and a battle tank together. Constructed of aluminum, it had two pontoons, each wrapped in metal treads, with a small deck and cab on top. Drew volunteered me to join its operator for the day. "You don't want to do torch; it's miserable in there," he said. "You can get a big-picture view on the Marsh Master. Got earplugs?"

The operator and I started scooping globs of glutinous mud away out of the vehicle's hitch to release an attached mower. I climbed up the pontoons and into the passenger seat of the cab. The operator turned the machine on. The diesel engine was so loud we shouted to hear another. Eventually I put my earplugs in, and we drove along in muted bubbles, communicating through hand gestures. We followed a trail along the edge of the woods to the lagoon, where we threw a suction hose into the water and started up the pump, filling the pontoons with a hundred gallons of water. Back in the woods, Drew and the crew, already laying down spots of fire, were barely visible as they trudged through a thick understory of reeds that now sounded like machine gun fire as they exploded with heat.

"We're going to hold up and let y'all get caught up, because down there you are just going to be in the smoke crap, but you're our first line of defense," Travis said to us over the radio.

We oriented the Marsh Master behind the ignitors, and I stood on the deck, bracing my legs, hose over my shoulder, spraying water, and putting out any spots as they blew across the line. Smoke began to flood from the woods. I was enveloped in a swirling white cloud, tears and snot streaming down my face. When the flames got too hot, I shielded my face with a shoulder.

Two crew members jumped in a motorized aluminum boat and began trawling the lagoon along the bottom of the unit, setting it on fire with torches. We turned the Marsh Master and drove to meet the boat where marsh and lagoon converged, the machine's treads barely visible above the squelching mud. I jumped off and grabbed a drip torch. With the Marsh Master behind me and the boat following in the creek, I began to light a head fire at the edge of the marsh and into the woods. The brown sludge gave off a salty, fermented scent and entombed my boots, forcing me to pull my leg out with every step. Behind me, the flashy reeds were erupting in fifteen-foot flames that twisted in the breeze. In the seconds it took to look behind me and check the fire behavior, the deeper my feet sank into the muck and the closer the flames moved toward me.

I realized with some alarm that my radio wasn't working: I could hear traffic but no one could hear me. How fast should I be laying down fire? I wondered. When I reached a bend, the flames fed into one, creating a shield of heat that was almost unbearable. I couldn't stand still; all I could do was move away from the heat faster, which meant putting more and more fire on the ground. I began to think through my options if a shift in the wind pushed the flames toward me instead of away. The only escape route, I figured, was to jump into the creek. When I came to the end of the line, I saw Travis in the woods. He had been keeping an eye on me the whole time. I tied in with the paved road and went over to him.

"Sorry about the radio," I said.

"I'm not worried," he said.

We walked a ways and sat down on the grassy ditch to watch the crew size up a burning snag. Travis drank from a big plastic bottle of Coke, and I chugged water from a metal canister. As the saw started up, we talked about this place. Travis had made sure I knew he was a redneck, but now he spoke like a naturalist. The marsh I had just burned was habitat for black rails, he explained. The rail was about as big as a sparrow, and it nested and foraged in the tall reeds. Black rails were notoriously secretive and hard to find. They were also disappearing. Two years before, the bird had been listed as threatened under the Endangered Species Act. Their decline seemed to be due to some combination of sea-level rise, fire suppression, increased salinity of their swampy habitat, and extreme weather events. According to Travis, the timing of our burn was designed to reduce the dead thatch and create regeneration in the rushes that would support the birds as they returned to nest.

"Burning helps with accretion," he said. "It stimulates the root growth."

"So, the cutoff for burning because of black rail nesting is March 15?" I asked.

"Yeah," he said. Then he sighed.

> Here's my deal on endangered species. It's good that we care and want to protect the species. But we're starting to pigeonhole ourselves to where we can't do *ecosystem* management. This is a unique system where we have longleaf pine directly adjacent to marsh. If you try to cut fire out, what you're doing is destroying an ecotone with some of the rarest plants and messing up the hydrology. Longleaf needs growing-season burns. Well, on these places that have marsh and longleaf, unless something changes, I'll never be able to do growing-season burns. [Red-cockaded woodpecker] requires growing-season burns. Then there's gopher frogs. All types of insects. Which do you pick?

If you managed for a single species, there was no good answer. "Our thought process is: restore the ecosystem. These animals are naturally adapted to this fire, so we mimic it as best we can. Putting fire back on the landscape is better than never doing anything at all."

I asked Travis where he had learned to burn. "I grew up watching my grandpa," he told me. State and federal agencies in North Carolina burned around 180,000 acres each year. Travis and his agency were burning as much as 50,000 of those acres. "I would love to see it go to 150,000," he said.

"What would it take to do more burning in North Carolina?" I asked.

"Some attitudes changing on smoke," he said. "Getting some legislation more like Florida and Georgia."

His crew eschewed government wildland fire qualifications for hands-on experience, but he felt they were often treated as outsiders as a result. "We get made fun of a little bit. We very rarely get called out on project fires," he said. "We don't even get invited to the trainings. And if we sign up, that's a day we can't be burning and cutting timber. But we're definitely all about prescribed fire."

"BACK-CUT! TREE FALLING!" shouted the sawyer from inside the woods. We could hear the sawyer whacking with the head of an axe at wedges she had put into her back-cut. The tree's trunk began to splinter and crack, and then I heard the whoosh and felt the thud in my chest as it hit the ground. The sawyer turned her chainsaw back on to cut the stump down.

"Tell her to leave that!" yelled Travis. He turned to me. "The reason I leave stumps high, or leave the whole thing if we can, is I've actually watched brown-head nuthatches create cavities two or three years in a row in those stumps as they rot and decay. Then I've watched chickadees use those cavities after. Nuthatches, part of their pair-bonding is creating cavities."

Drew emerged from the woods and walked over. "There anything more we can do, Travis?"

"I think we're good, man," he said. "If y'all want to head to Mayo's Seafood and get something to eat—I know it's gonna be a long ways for y'all to go home—I'm more than happy."

One of the crew members pulled up next to us in a truck. "It looks beautiful, man!" he said. "Y'all are some hardworking fools, you know that?"

Beautiful and Right

"I think they're fixin' to leave. I'm going to jump in with y'all," Travis told him.

We made our goodbyes. I mentioned to Travis I had never burned so many acres with so few people.

"Really?" Travis said, surprised. "We coulda done it with four."

ON MY LAST afternoon in Wilmington, I went to the warehouse to meet Drew and our module. They stood around a table that was like a wooden box on legs. Inside the box were a bunch of sand and an assortment of plastic children's toys: trucks and dozers, string, sticks, and miniature trees. Together we were going to re-create the topography of our burn in Pamlico County and talk through the operation.

"There's no risk to us right now," said the burn boss. "This is a great way, with retrospection, for us to pool experiences and see what we can draw out of it. We can bring up sensitive things with the understanding that everyone who burned together yesterday, everyone was there for the right reasons, everyone had good intent, everybody did the best they could. Nobody did the absolute perfect job."

As the burn boss was talking, Drew's phone rang and he stepped away to answer it. He returned to the table. "I've got kind of a curveball," he said. "I guess wherever the other crew is, they're having a hard time containing... something? They want us to load up and take off to help them. Anyone know where they're at by any chance?"

"Oh, they're by the Green Swamp somewhere," a crewmate said.

Drew thought for a moment. "We probably won't be back till after dark," he said. "I'm not exactly sure what's going on. Sounds like a lot of water and stirring dirt."

We drove west out of Wilmington and turned south onto a sandy access road with tall forest on both sides. We were about ten miles west of the Green Swamp, a seventeen-thousand-acre preserve of carnivorous plants, longleaf savannah, pond pine woodlands, and pocosin, and headed into state-managed timber and game lands. No one knew what lay up ahead. Could they really have lost control of their fire? In all his years burning in North Carolina, Drew had never had a burn escape to the point where operations went into suppression mode, let alone one that required calling in extra resources.

Finally, near a place called Second Cross Swamp, we began to see a milky pall of smoke hanging in the air. It got thicker, and we saw a couple of pickup trucks parked on the side of the road. To our right was the blackened edge of a unit the crew had burned that day. The problem was on the other side of the road. Somehow, the fire had jumped the road and was now actively burning in the ground; through the trees we caught glimpses of creeping flames and downed logs burning.

We parked and hoisted chainsaws and tools out of the truck bed and started to look for a way over a steep drainage ditch filled with water. About twenty yards down from the truck, we found a slender tree trunk that had fallen, providing a narrow footbridge. This tree was, it turned out, the reason we were there. Earlier in the day, someone had burned the grass in the ditch, assuming that the standing water would act as a fuel break. It did, but while the crew was busy mopping up somewhere else, the ditch fire had managed to crawl across the log. When the crew returned a couple of hours later, they had discovered a rapidly spreading forest fire.

I inched along the trunk sideways and then jumped the last couple of feet onto solid ground. As I looked around to get my bearings, I felt like I had stepped through a portal. It was almost magical, the way the forest had concealed itself from the road. We were in a hardwood bottom swamp, fecund with tangles of blaspheme vines and a labyrinth of enormous laurel oaks and swamp chestnut oaks.

Beautiful and Right

The canopy overhead was so thick I was submerged in a blue-green twilight.

The fire had essentially originated where I now stood and spread outward in the shape of a fan. I began to cut a path through the fan's center, stepping around flaming logs and ashy-white root pits while simultaneously ducking beneath and pushing away thorn-studded vines and weaving between the buttressed trunks of giant trees. I soon lost sight of everyone; I couldn't see even ten feet away through the dim light and dense vegetation. It was oddly quiet, like the air muffled sound rather than carried it. To orient myself, I had to stop and close my eyes to try to listen for the crashing of feet and murmur of voices in the distance.

Eventually, I found the perimeter of the fire and the crews digging line around it. A sawyer cut through the wall of vines, and another wrestled and bucked them aside, clearing a path for us to follow. It wasn't a line so much as a sinuous scratch that twisted and turned and looped about trees and logs and shrubs. I started digging and discovered that just inches beneath the surface of burning earth, the soil was sodden with water. I couldn't get my mind around it: we were in a burning swamp. My boots and pants were soon soaked and muddy. My rake was almost useless: with every strike it became snared in a web of roots. It was easier to scrape and pull at these capillaries with my hands. On my knees, digging into the earth, I had a sudden and strange sensation that I was *inside* something. I was surrounded by a seething mass of livingness. I felt the weird and sublime womb that is a forest, how it pulsed in a self-iterating expression of creation. I recognized the glorious whole and knew for sure that if I stopped to rest against a tree, it would absorb me into its plant flesh.

For the next couple of hours, we dug and cut and carried water and ran hose. I found an enormous laurel oak with a cavity at its base that fire had crawled up into. I reached into the tree's heart and scraped at the embers with my tool, sending showers of sparks downward. It was a waxing new moon, and when the last light left the sky, the darkness was blinding. We walked out of the forest, across the fallen trunk, and took off our dirty yellows in the light of the tailgates. Driving back, our truck hit a patch of mud, and the weight of the trailer caused the back wheels to sink. The truck, mired up to its running boards, couldn't go anywhere. As we waited for help to come, I realized that I didn't want to leave anyway. It would be nice, I thought, to fall sleep right there under a blanket woven of stars and smoke.

12 IN THE LAND OF THE WHITE DEER

IN THE VERY BEGINNING, WHEN EVERYTHING WAS DARKNESS AND SPACE, THE CREator known as *Wohpekumeu* decided to make the world. The first thing he created was soil, which he shook over the new world. Then he created the sun and the moon and all the rivers and oceans. The very first animal he created was the white deer, so pure and sacred even its hooves were colorless. Then the creator made the rest of the world's creatures and plants and the first human, a man. But the man was so lonely that the creator told him to blow his nose. When he did, the root of the angelica plant transformed into the first woman. The descendants of the first man and woman were the Indigenous tribes of northern California known as the Yurok, Karuk, and Hupa, and they still gather today to revere the first animal in a ceremony called the White Deerskin Dance.

On a sunny day I sat at a picnic table on the dance grounds of the Yurok village of Weitchpec. The branches of an enormous, gnarled bay tree shaded us, and its peppery leaves spiced the air. We were in a steep canyon, and the Klamath River flowed just a stone's throw away. At the edge of the picnic table stood Bob McConnell, a Yurok elder, in a baseball cap and short-sleeved denim shirt. Tall and slender, Bob rested his hands on a long walking stick made of cherry. "We're a long ways from nowhere," he joked. Not far from the Oregon border, the Yurok Reservation extends from the mouth of the Klamath River to its confluence with the Trinity River, and one mile on either side. But for the Yurok people, the largest tribe in California with six thousand enrolled members, the dozens of ancient villages that line the Klamath River and the surrounding mountain peaks and valleys are at the heart of the world.

Bob is a dance leader responsible for upholding the rituals of the White Deerskin Dance. In the Yurok language, such people are described as "doctors of the world" or those who "talk with the world." Bob explained, "The primary purpose of it is to give thanks and acknowledge what the creator provides for you. The Jump Dance is a dance that lays down the groundwork for the next two years. They are Healing the World ceremonies. Fixing the World, some say."

The Yurok view greed as a betrayal of the interdependence of life, a violation that can tip the world into imbalance. It was the *wo'gey*, the spirit beings, who instructed people to put the world back into balance through ritual dances. To prepare for the dances, debts are paid, grudges settled, and if someone in the community dies, a payment must be made to the family for the ceremony to proceed. "You're not supposed to kill a mosquito; you are supposed to leave death outside these ceremonies," explained Bob. "If a family member dies, people are expected to wait a year before they can partake in the ceremony." He turned and pointed toward the river where the White Deerskin Dance had concluded just a couple of weeks before. Traditionally, an odd number of dancers between nine and twenty-one gathered. That year, it had been thirteen, to match the number of annual moons. "Walk down this trail and

In the Land of the White Deer

there's a dance ground. There's a line of thirteen dancers, in recognition of the thirteen moons in the year, and singers in the middle. About a dozen feet in front of them is a circle where a fire is built. Once the fire is started, it's carried up the mountain, where another fire is built, and another down on the beach. All of the fires are lit from the same single fire."

Traditionally, a young girl of high birth tended the first fire for the duration of the ceremony, using its embers to burn *Lomatium californicum*, otherwise known as angelica root. "It smells really nice, and that smoke carries the prayers up to the creator," said Bob. "In the deerskin ceremony, when you are burning that fire, the temptation is to ask for cold and clear water. But you can't. You have to say, 'Thank you for cold and clear water.' It's a subtle difference."

Bob's family are regalia holders, meaning they possessed many of the sacred objects required for the high dances—woodpecker scalps, abalone shells, eagle feathers, bone whistles. Regalia embodied *mrwrsryrh*, or beauty, which is the fundamental quality of the world in Yurok spirituality, according to the anthropologist Thomas Buckley. Holding regalia was the highest possible form of wealth, but, as Buckley has written, it would be wrong to think of regalia as "objects." Regalia are people, "sentient beings (like deer) that have *wewecek*, 'spirit' or 'life' and that 'cry to dance.' As people, their 'purpose in life' is to get out of the baskets and boxes in which they're stored and dance to 'fix the world,' make the physical plane once again like the pure prairie of the sky, *wesʔonah*, from which they come." Regalia comes to people who are dedicated to fixing the world.

When the Yurok medicine woman Lucy Thompson published her book *To the American Indian* in 1916, she described the regalia worn during a jump dance at Pecwan in the late 1800s:

The wealthy ones that own lands, hunting territory, fishing places, slaves, flints, white deer skins, fisher skins, otter skins, silver gray fox skins and fine dresses made of dressed deerskins, with fringes of shells knotted and worked in the most beautiful styles, that clink and jingle as they walk and make one have a feeling of respect and admiration for them. Everyone will strain to look on this most pleasing sight, which can never leave one's memory that has seen it in its flowery days.

The anthropologist A. L. Kroeber called the religious ceremonies of the Yurok the "world renewal cult." By the time he wrote about them in the 1930s, he believed that the dances and many of northern California's Native traditions had already disappeared or were dying. It is true that the Jump and Deerskin Dances began to disappear from the villages of the Klamath River in the early twentieth century. But Kroeber (who was Omer Stewart's professor) died in 1960. He couldn't see the future. In 1984, fifty years after the Jump Dance disappeared from the village of Pecwan, it was resurrected for the first time and has been held every two years afterward. In the late 1990s, the community began to discuss reviving a high dance in Weitchpec. Bob worked for the Yurok tribe as a contact person with the company that owned the ceremonial site and intended to log it and the tribe. He also descended from the village of Wah sek, which historically participated in the ceremonies. The elder at the forefront of the revival effort said to Bob: "You are going to lead Wah sek." The first ceremony in nearly a century was held in 2000 and has been held every two years since.

"I can't say that I was born into it," said Bob. "When I was growing up, we didn't attend many of the ceremonies. A lot of them didn't happen. The one I'm a part of was not in existence at that time."

It requires a lot of preparation and thought to adorn the singers and dancers—who sometimes number as many as twenty-one people—with the necessary regalia, according to Bob. Most significant are the deerskins, which hang from long sticks carried by dancers. "Any pure color of deer can dance, but the white deer is the most prized," he said.

In 2010, the Smithsonian Institution returned 217 items of regalia—condor feathers, baskets, wolf headdresses, and white deerskins—that had been taken from the Yurok a century earlier. "These are our prayer items," Thomas O'Rourke, then the chairman of the tribal council, told a reporter. "They are not only symbols, but their spirit stays with them. They are alive. Bringing them home is like bringing home prisoners of war."

During the dance, four people wearing headbands of sea lion teeth and "blankets" (deer hides with hair left on them) blow on whistles made of the ulnas of blue herons or eagles and hold obsidian blades. As they move in front of the singers, they use the blades to cut through the invisible veil between this world and the spirit world. "That is one of the moments that I've felt the spirit world make things just stand still," said Bob.

> It's a good feeling. I don't know that I can describe it other than it's a high that you can't get anywhere else. It's a connection to another part of the world that I don't think many people understand. I certainly can't say that I understand it. But I know that there's a spiritual world that lives right alongside of us. I don't want to get too deep into this part of it, but during ceremonies you can feel those spirits surrounding you.... I'm convinced the ceremonies are very important in connecting the two worlds.

AMONG WILDLAND FIREFIGHTERS, the Klamath region had a reputation for misery. The heavily forested mountains are murderously steep and choked with poison oak. Each summer the region dries out just as cloud-to-ground lightning strikes are at their peak; a typical July might bring over fifteen thousand lightning strikes. The resulting wildfires often behave in extreme ways. They generate fierce amounts of smoke and pyroCbs whose lightning strikes create yet more fires. For all these reasons, there was a refrain among wildland firefighters: "Fuck the Klamath." The craft coffee company Hotshot Brewing even sold a T-shirt emblazoned with it.

But for residents of the region, the feeling toward wildland firefighters and the federal land-management agencies that employ them was mutual. Fire suppression crews were often seen as occupying colonial forces who took over roads and towns, whose burn-out operations created just as much damage as wildfire itself, and who felt little to no responsibility for the outcome of their tactics. "Locals here have seen more fire than most of the firefighters that come into the area," said Will Harling, director of the Mid Klamath Watershed Council and a lifelong resident of the Klamath. "There's PTSD from fire suppression. There's PTSD from watching the land grow sick around us and all the animals going away and humans going away from our communities because we haven't managed the land right. It's a lot deeper than just fear of fire. It's anger and resentment of a century-old campaign to take fire away from the people on the land. It's an ongoing occupation that gets highlighted every year when the incident management teams come in here and set up fire camps."

It was hard to imagine a place that had suffered more than the Klamath region from what some described as "government-subsidized capitalism"—the management of federal lands to benefit private commercial enterprises ranging from mining to logging to fisheries. The national forest reserves had been unilaterally carved out of Native land. The Klamath National Forest, Six Rivers National Forest, and Redwood National Park contained almost three million acres of land that occupied the majority of Karuk,

Yurok, and Hupa ancestral territory. Turning Native land into national forest meant that management decisions were taken out of Indigenous jurisdiction and given to federal agencies, where officials then signed contracts with private industries.

In the 1960s, the Forest Service began planning a federal highway through the center of what Yurok people called the High Country, a stretch of powerful medicine mountains where individuals went to fast, pray, and commune with *wo'gey*. The Forest Service slated sixty-seven thousand acres of the High Country's forest, containing hundreds of millions of board feet of timber, to be logged, but in order to create the hundreds of miles of necessary roads to get to the timber, it needed a highway. Yurok, Hupa, and Karuk people and allies fought what became known as the "GO-road" for years. They argued that the road threatened their right to practice their religion in the High Country, a right protected under the First Amendment. Others argued that the High Country could be saved if it were designated a federal Roadless Wilderness Area. But as the anthropologist Thomas Buckley wrote in his book *Standing Ground*, that suggestion highlighted the cultural basis of the conflict. "We want the sacred country preserved in any way possible," said one Karuk man. "But to do this by declaring it a 'wilderness' misses the point, and it would be better if some other means could be found. That country is not 'wild.' It was made perfectly by the Creator exactly as he wanted it to be there for us to use." In 1988, the case went to the US Supreme Court, where Justice Sandra Day O'Connor wrote the majority opinion siding with the Forest Service. (A section of the GO-road was never finished.) During that same period, the Forest Service had begun aerially spraying herbicides to kill broadleaf plants and make room for more conifers to grow for logging. Locals believed that the spraying led to a spate of bladder infections, skin infections, miscarriages, and molar pregnancies among women. According to the Karuk medicine woman and nurse Mavis McCovey, no living children were born in the area around the town of Orleans, California, between 1976 and 1978.

The consequences of these federal policies still reverberated. As I listened to Bob describe the White Deerskin Dance, I watched the flow of the Klamath River, glimmering in the sun, through the trees. Its waters are the lifeblood of the Yurok, but I had been told not to swim in it. The river was in the midst of a toxic blue-green algae bloom, a common occurrence. The blooms were created by the presence of six dams located on the river. Four of the dams were operated by PacifiCorp to generate hydropower, and two provided irrigation for farmers. The dams warmed the waters, creating conditions for the toxic algae. They also blocked hundreds of miles of habitat for salmon.

Bob had been born twelve years before the final dam was installed on the Trinity River. He recalled when he was younger watching as many as ten thousand salmon swim up the river to spawn in a single day. "These rivers use to be big and mighty rivers. There's still a lot of life left in them, but they're hurting really bad from a lack of water," he said. "I don't know how much longer this river can survive. This past year we had some really close calls and came close to having a terrible fish kill like we did in 2002." That year, thirty-four thousand Chinook salmon died in one week in the Lower Klamath River, the result of low flows from dams that warmed the temperatures so much it led to outbreaks of parasites and bacteria in the fish.

Of all the policies of the federal government, perhaps none was as existentially destructive to the region's first peoples as that of fire suppression. The Indigenous people were incendiaries. "We use fire to heat our homes, our food, to cleanse ourselves spiritually, to go into the sweat house to keep us renewed," said Frankie Myers, the vice chair of the Yurok Tribal Council. "It's the same in nature: it also needs to be cleansed; it needs to have fire on it." The Yurok lit frequent fires in the woodlands, meadows, grasslands, and forests, modifying fire regimes in order to support hunting grounds, acorn orchards, fishing, herb gathering, basketmaking, canoe building, and religious ceremony, all practices essential to their identity and survival as a people. As soon as European settlers arrived, the campaign to stop the Yurok from lighting fires began. In the 1920s, fire lookout towers were being constructed across the Klamath Mountains, according to the sociologist Kari Marie Norgaard, and were often built on sacred sites considered akin to altars. In 1926, a Yurok man by the name of Robert Spott told the Commonwealth Club of California that his people were "almost at the end of the road." For generations, the Yurok had picked berries, hunted, gathered acorns, burned, and fished in particular places. When they went to those places now, they encountered signs that said, "Keep Out." The fear of punishment from federal agencies lasted for generations. "When I was a young man growing up on the reservation, all of our land laid behind locked gates," Frankie Myers has said. "We'd have to break into our land in order to get into our prayer sites. In order to get to our gathering sites, we had to be outlaws." Elizabeth Azuzz, a Yurok tribal member of Karuk ancestry, recalled her father and grandfather telling her that if she saw a green or red truck while they were burning, it might be a governmental fire engine, and she should run away. Because at least "they wouldn't shoot a child, they wouldn't arrest a child," Elizabeth told me. "There were many times when the adults were arrested for doing what we do normally, which is burning our hunting ground or burning our gathering grounds."

Suppression changed the structure of the forest. Oak trees decreased by half while Douglas firs increased by 30 percent, crowding into what were once open prairies, savannahs, and hardwood groves that supported elk and deer and complex vegetation patterns. Despite extensive logging during the twentieth century, there are now three times more trees in the Six Rivers National Forest than in the nineteenth century. The result has been intense conflagrations that rip through the continuous fuels. At the dance grounds, Bob turned and pointed to the mountain slope behind him. "Green is a pretty nice color, but honestly there's too much green. Ninety percent of that bank right there is all [invasive] blackberry vines. The black locust that is here is invasive and is taking over," he said. "How much water would we be adding back into the river if we just took out the blackberries? Not to mention the thousands of acres of planted forests? We have more of a tree farm than a forest here."

Among California's Native tribes, there was once a widely shared belief that destructive wildfires were punishment for breaking religious laws. In Yurok culture the class of people known as Talth are women or men born of high marriages. Lucy Thompson, herself a member of this class, wrote that only the Talth knew the true name of the creator, and it was they who had kept the lomatium root to commemorate God's creation of women, planting it high up in the mountains. In her book, perhaps co-authored by her Christian husband, she described how one day the creator would punish human greed by sending a "great conflagration that will consume all the world in flames, and that its people will pass away. Over their ashes God will create another people, where they will build their stately mansions of the soul unto God. Over the ashes of the obliterated ages will prosper a new people with new governments and new laws, and the ages of peace and happiness will dawn again, shedding its radiance of glory over the entire world."

It was hard to be on the Klamath—where each summer people brace for the fire that could wipe out their communities—and not mull this prophecy. According to a study in the journal *Ecosphere*, between 2000 and 2018, over half a million acres burned in the region, compared to only two hundred thousand in the prior century of suppression. Studies have shown that global warming will bring a temperature increase of 3.5–4.5 degrees Fahrenheit in the next fifty years across the Klamath region, and that precipitation will be more variable, with extreme wet and dry years—a kind of "precipitation whiplash," as described in *The Klamath Mountains: A Natural History*. In the town of Orleans, the average temperature from 1931 to 2014 had already increased by 2 degrees Fahrenheit. On the first day of the White Deerskin Dance in 2022, the temperature had reached 107 degrees. On the last day, they couldn't hold the ceremonies because of torrential rain in what is typically the dry season. "Maybe climate change, huh?" asked Bob. "Maybe because there's too much brush here on the hillsides? Maybe it's because there hasn't been enough fire on the ground?" He paused. "No. It's not maybe."

THE CREWS I worked with in the Klamath region were made up of Indigenous fire practitioners and an assortment of wildland firefighters, landowners, and prescribed-burn association leaders. Unlike previous prescribed-fire-training exchanges I had attended, the groups were intentionally made up of people who possessed Native knowledge about fire and those who possessed Western knowledge about fire. But many of the non-Indigenous crew members were simply there to be what one Karuk woman described as "fire accomplices," giving our time, skill, and physical labor to the effort of restoring the health of the land, and strengthening the regional tribes' "fire sovereignty" and rights as cultural burners.

On the Yurok Reservation, this twice-annual gathering had been held for nine years. The idea for it had been hatched more than a decade earlier with the need to make a baby basket. Margo Robbins is a Yurok basket weaver, activist, and traditional-knowledge keeper. Her daughter was expecting twins, and she needed a basket for carrying the babies when they were born. When the anthropologist Thomas Buckley was interviewing Yurok speakers in the 1970s, many described a cultural belief that at a certain point of development in the womb, a child's fire or spark enters its heart and creates the individual's foundation or purpose. That is a person's spirit, and it comes directly from the creator. As the medicine woman Mavis McCovey explained in her book *Medicine Trails*, for the first ten days of life, newborns are wrapped tight, placed in a basket, and rocked to the sound of the mother mimicking the beating of the human heart. After a ceremony on the tenth day, the baby is rocked in the basket to a sound that mimics blood running through the baby's ear before birth. The practice serves to sustain connection between mother and child: "So they go on like that, and pretty soon the child is six or seven or eight years old. The mother is busy and the child doesn't feel well. All the mother has to do to calm the child down is to make that sound. She doesn't have to touch them. She doesn't have to get up and walk over to it. All she has to do is go 'Mmm, mmm, mmm' and the child calms down because it knows its mother is there."

Yurok baby baskets are woven from the long, thin shoots of hazels that only sprout from the shrub's roots the year after it has been burned. By the time Margo was to become a grandmother, it had been so long since fires had burned the hazel groves that there weren't any sticks to make a basket. "It was one of the driving points for me to really push hard to bring fire back to the land," Margo told the crew. "Not just for myself but because many basket weavers' practices were dying because we had no weaving materials." The tribe encountered numerous roadblocks to reintroducing fire. "Yurok people aren't ones to understand what [no] means," Bob explained. "We've done a lot of things that other tribes haven't because we will not take no for an answer." They also knew there was legal precedent for protecting a landowner's right to use fire as a management tool. Florida's 1990 Prescribed Burning Act was a landmark piece of legislation that protected responsible burners from civil liability. "In other states, like Florida in particular, they are way ahead of here," said Bob. "[Even] Georgia and Alabama have some pretty good programs for landowners to manage their land through the use of fire."

The Yurok knew they were approaching a point when it would be difficult if not impossible to reverse the course of ecological damage that had resulted from fire suppression. "Our culture is completely dependent upon fire," Margo has said. "Without fire, our culture will not survive. It's that simple." In 2013, Bob and Margo took the radical step of creating the Cultural Fire Management Council, an organization whose mission was to facilitate the practice of cultural burning. Soon after, they began holding training events and inviting people to come and burn on the Yurok Reservation. Two years later, they joined forces with the neighboring Hoopa and Karuk Tribes in collaboration with the Nature Conservancy's Indigenous Peoples Burning Network, which supports and shares resources for cultural burning. In the years that followed, the network expanded to include New Mexico's pueblos, the Klamath Tribes in Oregon, the Leech Lake Band of Ojibwe in Minnesota, the Washoe Tribe in California and Nevada, and the Alabama-Coushatta Tribe in Texas.

Thanks in part to these efforts, in 2022 California passed its own landmark pieces of legislation that limited the liability for tribal members and private citizens conducting controlled burns. One of the bills was AB 642, which defined a "cultural fire practitioner" as "a person associated with a California Native American tribe or tribal organization with experience in burning to meet cultural goals or objectives, including for subsistence, ceremonial activities, biodiversity, or other benefits." It also directed land-management agencies to increase the amount of cultural burning in the state. One of the immediate effects of the law was that for the first time since 1850, the Yurok and other tribes in California could tend to their land without the fear of going to prison.

Arson was a complex issue on the Yurok Reservation. Elizabeth Azuzz, the secretary of the Cultural Fire Management Council, told me that several of her own family members had been sent to prison for setting fires. One of them was burning brush on family property and is now barred from touching flint, matches, a lighter, anything that could cause a spark. "His life is scarred because he didn't believe he had to follow procedures, policies, listen to anyone other than our elders," she explained. "This man in his late forties literally has to have a family member come over to light his barbecue. He cannot touch it. He can't even build his own home fire. He did a great deal of prison time for helping his elder burn brush on their family property."

In the past, local resentment toward the system that restricts and penalizes Native traditions manifested in more arson fires. Along a single twenty-mile corridor of Highway 169, a road that runs alongside the Klamath River from the village of Pecwan to Weitchpec, there were as many as eighty cases of arson a year. (Bob told me that the Hoopa and Yurok Reservations were once first and second in the nation for the number of arsons.) "That corridor is in the densest interior of our reservation, and most of the population that lives down in there have been there for generations," said Elizabeth. "They still believe as we all do that it is our God-given right to restore and maintain the land with fire. They have a huge aversion to permits or any government agency telling them what to do."

Early on, the Cultural Fire Management Council received a grant with CAL Fire, California's firefighting agency, to reduce the density of forest along roadsides and to create defensible space from wildfires around homes. In subsequent years, the number of arsons on the reservation plummeted. The more fire that was returned to the reservation, it seemed, the less likely people were to start unsanctioned fires. "I've actually made some arsonists participate in our training," said Elizabeth. "They fought me all the way, but in the end one of the young men stood up in front of the entire room of people and broke down as he apologized for everything that he was doing. He didn't realize how much of an issue he was causing for everyone else while he was dealing with his anger and resentment."

Nine years on, the Cultural Fire Management Council had burned thousands of acres. Bob, its former executive director, called it "growing the forest backwards." The goal was not to cultivate more of every type of ecosystem; the goal was to use fire to carefully create a mosaic of various successional stages, from grass to savannah to forest. "You can't sit back and just let the forest grow; you have to get out there and you have to manage it," he said. "Fire was *the* tool that our ancestors utilized the most. It was the sharpest tool in the cabinet then, and it is now."

As the crew left the dance grounds in Weitchpec, Bob instructed us. "You folks learn how to take care of fire." I didn't know what he meant, but I thought about it in the days and weeks to come. We often burned in places that hadn't experienced a fire for over a hundred years, where big old black oaks and madrones—evidence of a once-open woodland—were drowning in jungles of Douglas fir. I lightly scraped leaves and duff from the bases of three-hundred-year-old tanoaks to better protect them from the heat of the flames, careful to leave the white tangles of mycorrhiza intact. I flushed deer while dragging torch in some of the few remaining white oak groves in the region and dug smoldering root pits in the dark, trying to learn how to take care of fire.

ON THE YUROK Reservation, I slept near a creek in a shaded community park known as Acorn Flat. It was the season when the trees began to drop their nuts, and all night I could hear the smack of acorns from the mature tanoak trees that towered over my tent hitting the ground, where they would lie until people came to gather them. We met for meals and operational briefings at the Woo-mehl Community Center, a gathering space on top of a hill that housed a kitchen, outdoor pavilion, fish smokehouse, and basketball court. In the early mornings, I stood on the court and looked at the too-verdant mountain slopes across the river that were shrouded in fog. I was in Bigfoot country, again. Just fifteen miles upriver was a tributary of the Klamath called Bluff Creek, where in 1967 Roger Patterson and Bob Gimlin were riding on horseback when they saw a hairy, bipedal creature standing near a logjam. Patterson grabbed his 16mm camera and pursued the creature. The resulting film is less than a minute long but shows what he claimed was a Bigfoot turning its head several times toward him before disappearing into the trees. Frame 352, the one where the creature looks directly into the camera, has been a source of fascination and speculation ever since. "I've heard 'em before when we were teenagers. We went camping up in the mountains with my friend's mom for like two weeks," said Margo. "Around here, Bigfoot looks out for women. So, we don't need to worry."

Margo, a mother of five and grandmother of nine, had long silver hair and the traditional facial tattoos once universally worn by Yurok women: vertical lines of dark ink adorned her chin. The walls of her home high above the Klamath River were covered in family photographs and sports trophies. She was quick to laugh. Her favorite herb was lemon balm, a plant that she said makes one's heart and mind merry. But Margo loved many plants, and when she walked through the woods, she seemed to see plants in multiple dimensions: history, spirit, medicine, family, and survival. "The *wo'gey* spirit beings lived here before us," she explained.

> It's not by chance that we know which plants to use to make baskets or how to use them. They taught us everything we need to know to live here in a good way. All of the things that our baskets

are made of were once spirit beings. All of the animals, the trees, the water, the rocks. They came with the agreement that they would take care of us, and they took physical form to do that. So when we go out and put fire on the land, we are honoring an agreement to take care of them, physically and spiritually.

Margo pointed to huckleberry, alder, maple, tanoak, pepperwood, and hazel and explained their various uses for colds and arthritis, clothing and food. Standing in front of a mature hazel bush, she told us how the woody stems were used in the traditional game called sticks that young boys and men play. It is both a test of physical strength and a form of prayer. Two teams of three, she explained, each tried to fling a wooden toggle downfield with their sticks as the other team tackled and held them on the ground and tried to get the toggle to the opposite end of the field. (Someone else described it to me as "a polo game with, like, three wrestling matches going on at once.") The Yurok are known for being champion wrestlers, regularly winning major competitions and sending members to professional mixed martial arts leagues; Margo's youngest daughter was a three-time state wrestling champion.

But what the Yurok are world renowned for is basketry. Twining weft strands around a vertical warp in browns, reds, yellows, blacks, and golds, Yurok women practice the art using materials like pine, ferns, willow, alder, serviceberry, spruce, bear grass, and hazel. The resulting baskets, some as tall as an adult and others dainty enough to be worn as adornment, are beautifully symmetrical, patterned with parallelograms, isosceles triangles, and zigzags. In the past, baskets that introduced new colors or motifs were considered "against the law," while those that upheld traditional patterns were considered "put together right." Women weave baskets for ceremony, drying tobacco, gambling games, fishing, collecting firewood, and cradling babies. When it comes to acorns, staple of

Yurok diets, there are baskets for gathering, storing, grinding, sifting, cooking, serving, and eating them.

The ethnoecologist M. Kat Anderson has estimated that a single Klamath village required hundreds of thousands of straight branches for creating the baskets necessary for pre-colonial everyday life. The same bush that Margo pointed to as having good material for the game of sticks was useless to her as a basket weaver, its limbs not small, straight, or limber enough. Only frequent burning produced the right materials. For a long time, fire suppression made them impossible to find, so it was impossible to teach the next generation how to weave. As Sharon Levy explained in an article for *BioScience*, by the 1970s, tribal women were forced to scour the places where logging operations had taken place and slash was burned in the hopes of finding good hazel and bear grass. The Karuk basket weaver Kathy McCovey told Levy, "When I lived up the Klamath, a lot of times you'd hear the engine roaring out of Orleans to put out a forest fire. My husband would say, 'That's my grandma out burning.' The little old ladies would walk along and see a spot that looked right and set a fire. It's a real pain in the butt to go through all the paperwork required for an approved, prescribed fire. It's easier to just drop a match. But now we have such a fuel load that the fires burn too hot."

Margo described the colonization of Yurok land as the "rape" of the forest. "They looked at it as a way to make money, not something that we are dependent on," she said. The lack of fire created a forest that today looks like a hoarder's house: "You can't take a bath, it's full of stuff. You can't use the sink," she said. "There's little trails to walk through the stuff." To begin reintroducing fire to these places often required what some described as "corrective" fire: controlled burns that burned hot enough to meaningfully reduce fuels.

Margo still had the baskets that her own grandmother had made. She showed the crew one of them back at the community center. The black, she said, came from maidenhead ferns, the white from bear grass, and the red from dye made of alder bark on woodwardia fern; all the fibers twined around willow. It was beautiful but, Margo said regretfully, not her grandmother's finest work; she had made it at the age of ninety. Willow played an important role in the creation story of fire. As Margo held her grandmother's basket in her hand, she shared this story in lyric detail, barely pausing to take a breath.

Way back when, she said, fire was hoarded by three old ladies, who lived in the sky and kept it in a basket hanging on their wall. The animals, who were in spirit form at that time, gathered together and made a plan to steal it from the ladies. *Wohpekumeu* went up to their sweathouse and was invited to spend the night. When they were sleeping, he stole the fire and took off running. It got cold in the sweathouse, and the old ladies woke up and started chasing him. *Wohpekumeu* ran as fast as he could and passed the fire off to the bear, who raced up a hill and gave it to the inchworm, who stretched itself all the way across to the next ridge. "But those ladies, man, they was fast, and they wanted that fire and they wanted it bad and they were gonna get it," said Margo. "When he got over there, inchworm passed it to eagle, and eagle grabbed it and was flying as fast as he could with those ladies behind him, and he passed it to mountain lion." Finally, frog, who was sitting at the edge of the river, took the fire into his mouth and went under the water and just waited. "Those old ladies got tired and went home," said Margo, "and frog came out and spits the fire into the willow."

She picked up a basket used for leeching acorn flour. "These materials you see are fire dependent," she said. "The work that you guys are doing when you come is critically important to the continuance of our culture, and this is just one of the things, basketry.

We're also creating habitat for the deer. Every place we go and burn, the deer are coming, and our hunters don't have to go off the rez anymore and risk huge fines for hunting off the rez."

"We now have elk back in the area from all the fires that we've been doing," added Elizabeth from the other side of the room. The crew whistled and clapped.

"You know, in a way," said Margo, laughing, "we're like those old-time spirit beings that stole fire from the old ladies in the sky. But we're just taking it from the government."

13 A PRAYER WE MAKE TOGETHER

"BRING YOUR GOAT LEGS TOMORROW," SOMEONE TOLD THE CREW ONE NIGHT. "THAT place is *steep*." We drove out to a unit called Hee' Mehl in the morning. I stood at the top of the handline and looked down. I burst out laughing. It was treacherous—not just steep but almost impassable. In places, the angle of the two-foot-wide path of mineral soil dug into the earth was greater than forty-five degrees. Sometimes it just dropped away completely. "Legit cliffs," somebody said and shook their head. "You'll slide on your ass for a hundred feet if you fall." To prevent the crew from wasting energy or risking injury trying to hike up the handline, the entire day's operation was designed so that we would all progress in one direction only—downward. We staggered our squads, each one monitoring a section of the "backdoor" for any spot fires or slopovers and moving down the mountain throughout the course of the day.

Hee' Mehl was 148 acres. Part of it had once been open prairie, and the crew had burned some of it the previous week. It burned hot, and the fire went into the crowns of the encroaching Douglas firs. Everyone cheered as the firs torched one by one like matchsticks. "You guys definitely killed some trees! It looks fucking sick," a crewmate said admiringly as we drove past.

We stood in a circle with our packs and tools for the operational briefing. Elizabeth Azuzz stood outside the circle, holding an angelica root in her hand and lighting it until an ember glowed. As we stood listening to the weather report and to discussions about fine and dead fuel moisture, communication plans, and medevac routes, she moved around and washed each of us in the celery-scented smoke of *walth-pey*, the root that had transformed into the first woman at the beginning of creation. Some took their hands and swept the offered smoke over their heads and bodies. Nearby, Margo Robbins prepared a bed of dried sticks, and at the end of the briefing lit a torch made from dried wormwood. Then Elizabeth lit her wormwood torch. "I used to do this by myself in the woods until Margo caught me," she laughed. Together, the two women began to pray, letting the plants and animals know that we were there to take care of them and asking for the crew to be protected. "We are co-managers of this land with creator," explained Margo. "So we want to make sure creator is right here with us." Elizabeth spoke:

> Grandfather. Grandmother. See us here, your children. Guide our hands as the seven generations of the future ask for guidance from the seven generations of the past. Protect those who came here to help us restore our land. Protect the one-legged tree people, the four-legged brothers and sisters, the winged brothers and sisters. Help us to restore our food and medicines. Help us to balance our ecosystem. *Wok-hlew'. Wok-hlew'. Wok-hlew'.*

On certain parts of Hee' Mehl, there was nothing else to do but sit on your butt and use your feet as a brake as you slid downward. Climbing back up even a few feet involved slamming the toe of your boot into the hard earth for purchase and grabbing any little plant nearby, hoping its roots would hold as you hoisted yourself up. My squad kept pace with an ignitions team that set fire to the interior of the unit, while a single ignitor lit off of the handline. I watched as she coasted down the mountainside at a perpendicular angle, an avalanche of rocks and leaves and dirt following her, clutching the drip torch with one arm while holding the other arm out in front for balance. Later she said it had felt like surfing a wave. It gave her blisters on the tops and bottoms of her feet.

We burned seventy acres that day. "Love it when the smoke is there," said the holding boss admiringly. "You got the rays coming down. Makes it feel like there's a supreme being here." As the sun set, the forest was cast in a violet haze and became ghostly, quiet, and ashen. We left the handline by jumping down the cutbank, our boots slamming onto the paved road. We walked in darkness toward the flash of engine lights and the ambient glow of the forest on fire. The crew gathered round the vehicles and watched showers of glittering embers tumbling down an exposed rock face, our own firefall. The next day, as we patrolled the road for fallen rocks and snags, a passing car slowed down, and a man leaned out the window, shouting at us with a big smile, "It looks beautiful! Thank you!"

ALONG THE TWISTING Highway 169, we burned an eleven-acre unit called Pasko one day. The forest was a vicious knot, the oaks and madrones and pepperwood trees choked by an overgrown understory of hazel and blackberry and young firs. I had gotten my certification as a fire effects monitor, and my role for the day was to make fire-behavior observations, noting the vegetation and how different patterns of ignition or fire intensity affected it, particularly the patches of mature and overgrown hazel. I worked with a trainee named Thea Maria Carlson. She was tall and pretty with hazel eyes and short, dark hair that she covered with a frayed baseball cap whose rim was emblazoned with the word "FEMINIST." As we hiked together, Thea told me that she lived in an intentional community near Santa Rosa. Located on four hundred acres, it was rooted in Quaker traditions and the principles of environmental stewardship. The land had gardens, orchards, grasslands, chaparral, woodlands, and farm animals, as well as homes, a community building, a barn, and a workshop. But in the years after Thea had moved there, nearby wildfires had forced the community to evacuate their land again and again. To mitigate the seasonal threat, the community decided to do an experimental six-acre prescribed burn and then voted to carry out even more prescribed burns in 2020. That September, however, the Glass Fire started in Napa County. In a day, it grew by over thirty thousand acres at a rate of an acre every five seconds. Ninety-five percent of the community's land burned, and twelve of its thirteen homes were destroyed. Pretty much the only place that did not burn was the six acres they had treated with prescribed fire. The Glass Fire was why Thea had come to the Klamath. She signed up to train as a wildland firefighter a week after it had been contained. The only path she saw to helping her community survive the future was by becoming a burn boss. "I had to embrace fire even though I was terrified of it," she said.

To make fire-behavior observations, Thea and I followed the firing boss, a guy named Raven, around the interior of the unit. Heavyset with tightly shorn black hair, Raven was a member of the Modoc Tribe and a veteran firefighter. He was quick despite his size and completely unafraid of intense heat or poison oak, which we were now wading through. Raven ran an engine with a private contractor in Oregon, but his other job, he told us, was as crew boss to a ten-person fire crew made up of young Indigenous men that called themselves the Wagon Burners. "I told them they were gonna have to change it," he said to me, shaking his head and smirking, "but they won't do it." Raven lit strips of fire that ran upslope, wild and hungry, while Thea and I slipped and crawled up and down ditches and through dense trees to stay below and ahead of his flames. Over the radio, Raven instructed the other ignitors, making sure he wasn't sending fire toward them and tracking their movements in relationship to us. As we reached the far side of the unit, one of the ignitors appeared in the trees, a Yurok man in his early twenties. Sweat poured down his face, and he sat on the slope leaning against his pack, panting for air. "Let's get experimental," Raven told him. "Let's get some bigger strips."

They headed back into the trees, and Thea and I hiked toward the handline. Just as we reached the bottom, however, a call came in over the radio. There was so much smoke socking in the handline that they needed replacements so the holding crew could get fresh air. Ideally, prescribed fires produce enough thermal energy that the buoyancy lifts the smoke particles high above the fire and carries them away at the atmosphere's mixing height, the number of feet above ground that smoke can be dispersed by wind turbulence and diffusion. I had seen ignitions teams coordinate with one another to create firing patterns that increased the heat and energy of the fire in order to lift smoke up and over the crew. But the forest we were in was so choked with brush and trees that everything was smoldering and smoking long after the flames moved through. Meanwhile, the afternoon's upriver canyon winds pushed the thick smoke directly over the handline. Thea and I looked at each other. "Are you good to go back up?" I asked. "Yep," she said. I called the holding boss on the radio and told him we were on our way.

We slung our packs off and got out goggles and bandannas, tying them around our faces. We started trudging back up the steep slope. Within a few hundred feet, our visibility was reduced to a few feet. The particles seeped around the edges of my goggles, stinging my eyes until tears streamed down my face. I tried to take even breaths through my nose, but the exertion of the hike made me want to gasp for air, which I did, turning my face away from the fire and gulping in vain. I fought the sensation of being smothered. At the top, we relieved two crew members and spaced out to look for spots across the line in the grass or the trees. Too far away from each other to talk, we stood in spheres of silence as the afternoon stretched on and the smoke washed over us. I watched as the fire chasers, small brown beetles that smell smoke through their antennae and lay their eggs in burnt wood, arrived, flitting around the charred logs and trees. At my feet, I found the azure feather of a scrub jay and picked it up with my grimy fingers. One of my fire pack's pockets was dedicated to discoveries like this: interesting seeds, leaves, bones, feathers, and scraps I found in the woods. A grey bird, maybe a vireo, landed on a nearby tree branch and perched for a while, a shadow in the smoke, watching the fire beetles too. At times, the intensity of the smoke lessened, and I looked for pockets of air so I could take off the bandanna for a moment and blow out the endless snot from my nose.

Around six o'clock, I sat on a tree stump to rest. I felt gloomy. It had taken over six hours to burn just eleven acres of forest. It was a speck amidst millions of acres that also needed fire. The reality of the work felt overwhelming: the people, planning, preparation, money, vehicles, water, fuel, and time. Meanwhile, the time window for reintroducing disturbance seemed to be narrowing every year as the meteorological thresholds—air temperature, relative humidity, wind speed—used to safely conduct prescribed burns became constrained by extreme fluctuations in the climate. In 2022, a study in *Science of the Total Environment* reported California's burn windows in the winter and spring have been decreasing by one day each year since the 1980s. Other studies have found that climate change will reduce opportunities for prescribed fire in the Southeast, while windows across the Great Plains could shift from spring to fall. The studies all seemed to point to the same conclusion: burning in the future will depend on people who are ready to exploit any and every opportunity as windows open and close with less and less predictability.

THAT EVENING, FRANK Kanawha Lake ate dinner with the crew. A research ecologist for the Forest Service's Pacific Southwest Research Station, he is a Karuk descendent with family that are both Yurok and Karuk. His career has straddled traditional and Western knowledge; he has published dozens of peer-reviewed journal articles focused on cultural burning, climate change, and forest health aimed at establishing the credibility of Indigenous land-stewardship practices within the scientific community. Frank had an intense, rapid-fire way of speaking that gave him an aura of missionary purpose. That night he talked to the crew about the lessons he had learned as a child from his elders, who had introduced him to the world of thunder beings, sugar pine spirit persons, and lighting medicine in the sacred places on the Klamath. "Everything here is your relation," he was told. "Every spirit manifested in the physical form has a teaching. What you need to do is learn about that teaching in a respectful way. You need to be open to learn, open to watch and see what's happening."

Later in life, Frank told us, he recognized this epistemological system as the foundation of a powerful spirituality. "Our medicine people, the High Men in Karuk and Yurok culture, were the ones who observed nature. I believe they worked with those spirits to be able to control things," he said. Take weather, for example. Indigenous people compiled information about wind patterns, atmospheric moisture, seasonality. "It's not an accident that people had a song for fog, a song for rain, a song for fire behaving the way you want it to," he said. "These things might be superstitious, but there was an acute understanding of that place, weather, environment, and topography. Those are the things that we're trying to revive."

Frank left rural northwestern California to get a bachelor of science from the University of California, Davis, and moved to Hoopa after graduation to work as a fish habitat biologist. It was 1999, and that fall he witnessed one of the first megafires in the

Klamath-Trinity region. Called the Megram Fire, it ran six miles in a day through Forest Service lands and onto the Hoopa Valley Indian Reservation, creating a massive smoke column. When the rains came in November and put it out, the fire had burned over 125,000 acres. After that, Frank figured that if he was really going to protect watersheds and restore fish, he needed to understand fire. And in order to understand fire, he needed to better understand the cultural history of his own people. Frank set out to talk with his elders and undertake the studies that would illustrate the validity of their traditional knowledge. "If I helped create the best-available science with a team of graduate students and top professionals," he said, "then they can't refute that. I can own my own cultural subjectivity, but the objectivity of the study design and the findings will speak for themselves."

In the world of the Klamath River Tribes, fire was energy, spirit, tool, and medicine. People learned how to live with fire and understood how it connected to everything else: baskets, food, water, springs, and salmon. Now Frank saw that many of the ecological theories he had learned as a scientist had analogous cultural teachings, but the Klamath River Tribes described them through a spiritual or metaphysical lens. "It's not by accident that the most sacred places are the most biologically diverse," Frank said. "In this area, what manages that diversity is fire. The White Deerskin Dance and Jump Dance, all of the regalia. It goes back to fire." The Yurok attributed the biodiversity of the land to their spiritual stewardship rather than a natural state. For instance, the height-to-crown ratio of the oldest, most productive oak or tanoak trees wasn't the result of random lightning strikes clearing away competition; it was the result of people burning low-intensity, frequent fires every three years. The abundance of iris grass in open woodlands wasn't accidental; people needed its leaf fibers for making nets and cordage, catching eels and salmon, and for making snares and dance regalia. Frank had

A Prayer We Make Together

attended the White Deerskin Dance held in Weitchpec just a couple of weeks before. He saw the regalia—the woodpecker crests, buckskins, and abalone shells—as indicators of ecological health and abundance.

"How do you get a white deer, an albino deer?" he asked us. "It's one in ten thousand. The more healthy a population you have, the more likely you're going to have an albino." There was no way to support such a large population of deer without frequent fire. When Frank hunted deer, his grandfather taught him to look carefully at the animal's insides. How many parasites were there? How much fat surrounded the organs, ribs, and hindquarters? A healthy deer was one that had access to springs refreshed with minerals from the fire; ash they could roll around in to get rid of ticks and mites; charcoal that cleansed their gut parasites; fresh grass that grew after the fire. If his grandfather gutted a deer and found it riddled with parasites, that was an indicator there hadn't been enough fire on the ground. Woodpeckers were indicators of a fertile forest, too. Known as a keystone species in biology, they supported dozens of other birds through their foraging and nesting. They were also the first to colonize newly burned forests. "All those woodpeckers you need [for dancing], each of those big pileated woodpeckers, there's a creation account about that too," Frank said. By understanding the relationships between environment, animals, and fire, people were able to produce quality materials for ceremony.

Frank liked to say that, for the Klamath River Tribes, the mountains had been and is hardware store, supermarket, pharmacy, and church. The forest produced by suppression, on the other hand, was what he called a genocide forest: high fuel-loading, closed canopy, no light, fewer species. "You can't live in that forest." Frank viewed first-entry burns like the one we had done that day on the Pasko Unit as the difficult but necessary first step in a long process toward Yurok families being able to burn their own land again. It might take three or four burns to achieve that goal. "It's going to take that initial entry with the holding crew so that families can take over in maybe ten to twenty years," he said. "The first time you come here, you might be holding. The next time you come back, you might be picking acorns or basketry material. So you can be like, 'I was burning here a couple years ago, and now I'm here gathering, and I remember how this was.'" In this future, with enough prayer, intention, and song, Frank envisioned that the spirits would even begin to talk back to people. They would know that they were being cared for and had a purpose again. "We have a long way to go to thinking about Indigenous spiritual understandings of place," he said. "Energy. Power. Medicine."

I HAD NEVER burned with so many women before as I did in the Klamath. On the Yurok Reservation, of the roughly forty-five crew members, sixteen of us were women. Three of the leadership roles—burn boss, agency administrator, and planning section chief—were filled by women, which was rare. Being around other women on a fireline was heartening, but for reasons that I couldn't always articulate. We seemed to seek each other out, driven by curiosity, the question always: How did you get here? Our ages spanned from twenties to sixties. We were firefighters, botanists, students, land stewards, lawyers, singers in punk bands, mothers, and grandmothers. I watched us fell and buck trees, drag torches across mountains, drive engines, and command sophisticated burn operations. It gave me a sense of intense pride that I physically felt in my chest. We did the work. We smashed.

Margo and Elizabeth led us by example. They were as quick to laugh or hug or give thanks as they were to assert their experience and authority. They spoke with directness and without apology. It was the same quality that I had recognized in Dan Kelleher and

Patty Carrick: they were so fully themselves. Sometimes after dinner, I would overhear the leadership team discussing the next day's burn, and the folks from state and federal agencies would doubt whether it could be pulled off. Margo and Elizabeth would confidently push back. "When you allow people who are suppression-minded to burn, they panic," Elizabeth told me. "A lot of time, I'm just like, 'You need to stand here and watch it, make sure it stays where it's supposed to be and that it does the job it's supposed to do.' As I always say, we don't fight family. So we don't fight fire." Later, Margo griped, "With all these megafires, they finally realize, 'Oh, maybe we should ask these Indigenous peoples how to take care of the forest.' We took care of it for ten thousand years."

During my second week on the Klamath, I worked with a different crew, upriver. The burn boss was Kelly Martin, a wildland firefighter with four decades of experience in suppression and prescribed fire. She had recently been appointed to the Wildland Fire Mitigation and Management Commission, which, as part of the Bipartisan Infrastructure Law, was tasked with creating recommendations to Congress for federal wildfire policy. One afternoon we sat in the shade of some oak trees outside the Karuk Tribe Department of Natural Resources to talk. Now in her fifties, Kelly said that after a long career in federal jobs, she felt a personal responsibility to give her leadership skills and expertise to the cause of good fire. "Suppression almost feels like the Vietnam or Korean War," she said. "Who are we fighting? And why? I have purposefully given up working on large incident management teams on wildfire because this to me is the future," she said, nodding at our surroundings on Karuk ancestral land. "This is our opportunity to lessen large megafires into the future."

We talked about Yosemite, where Kelly had worked as the chief of fire and aviation for ten years. "I was forty-two when I got there, and I remember thinking, 'OK, I've got a lot of time I can spend here to really make a difference,'" she told me. "I felt like I was in a place that was accepting of fire, and when I would go out there, I would have this really powerful experience with fire. Fire is a living force." Yosemite was a place of profound realization. Kelly had a sense of being in conversation with the landscape, which was telling her, *I'm not getting water and nutrients. I need fire.* "It was almost, like, this epiphany of how, through space and time, the landscape needs human help," she said. "Humans created this condition, and humans can step in to remediate it. We have no time to waste."

Toward the end of her tenure at Yosemite, Kelly decided to testify as a whistleblower before the US House Committee on Oversight and Government Reform. Her testimony included descriptions of pervasive gender bias, workplace hostility, and episodes of sexual harassment throughout her career with federal agencies. It led to a wave of retirements among career national park administrators. Her decision to testify had led to more activism and advocacy, and she became the president of Grassroots Wildland Firefighters, a nonprofit agency that advocates for policy reforms around mental health, pay, and classification for the country's wildland firefighters. But her passion was supporting women in prescribed fire. A couple of weeks earlier, Kelly had been the burn boss for a prescribed-fire exchange that brought together Indigenous women from North America and Australia representing two dozen different tribes.

Advocating for good fire and advocating for women were intertwined. Kelly described the culture of fire suppression as a chauvinistic, "militaristic view of wildfire as a war to be won and an enemy to be fought." That led to a collective psyche of dominance that shaped leadership, teams, and individuals, and to a permissiveness around abuse of land and people and women. Kelly believed the success of the good-fire revolution was dependent on the ability to change this culture. "We all have different backgrounds, all of us are diverse in different ways. The integration and synergy that needs to be created when we go into prescribed burning is essential," she explained. "If we are not in alignment with each other, the way we treat each other, the way we communicate, the way we behave, we are a liability rather than an asset."

Kelly foresaw a long road for the good-fire movement. Billions of dollars and a military-like structure had been committed to putting fires out. "You have lobbyists for retardant planes. It's a suppression-industrial complex. It's extremely harmful in so many ways," she said. "There's a place and a time for protection of life and property. But there should be a *greater* space for the use of fire on the landscape. It does feel like a David and Goliath situation with the prescribed-fire folks." What, I asked her, did she think would tip the scales? Was it a change in leadership at the federal agencies? The resentment of citizens realizing their tax dollars have been used to create the fire paradox now threatening their communities? New media narratives around wildfires? None of these was a magic bullet, said Kelly. But a good place to begin was to allow federal lands to be managed by America's tribes. "Suppression has put many Native cultures into a position of poverty, lack of food sovereignty, loss of hunting and gathering grounds, [and] traditional practices," she said. "If tribes said, 'We want to take back Indigenous burning on national forest land,' there shouldn't be anything to stop them."

The Ojibwe Indian scholar and novelist David Treuer has argued that America's national parks should be returned to its tribes. "For Native Americans, there can be no better remedy for the theft of land than land," he wrote in an essay for *The Atlantic*. "And for us, no lands are as spiritually significant as the national parks. They should be returned to us. Indians should tend—and protect and preserve—these favored gardens again."

Tribes are already on the forefront of prescribed-fire use. Between 1998 and 2018, the Bureau of Indian Affairs was the only federal agency to increase the use of prescribed fire, and some 50 to 80 percent of its fire suppression budget is given to prescribed fire. Since 2015, the BIA has allocated millions of dollars for tribes to reduce fuels and conduct prescribed burns on landscapes within and adjacent to reserved treaty-rights land, the ancestral places that were used for ceremony, hunting, and gathering and are often located on federal lands. A five-hundred-acre prescribed burn conducted by the Pueblo of Tesuque under this Reserved Treaty Rights Land program, for example, was credited with preventing the 2020 Medio Fire from burning through the city of Santa Fe's watershed.

Although Native traditions of fire use have been disrupted, some even lost, many Indigenous communities are striving to maintain or rebuild this knowledge. The impact of their efforts is significant and transcends tribal boundaries. As Frank Kanawha Lake said, "What we do in our Indigenous community's interest is also going to make society a cobeneficiary. Fire risk reduction. Lower suppression costs. More biodiversity attributes. Water services." For these reasons and others, Indigenous land rights are increasingly recognized as a critical aspect of climate action. Eighteen percent of the planet's land is under legally recognized Indigenous stewardship and contains 80 percent of its biodiversity. Seventeen percent of the world's remaining forest carbon is in land stewarded by Indigenous communities. In recent years, the Yurok have increased the size of their land base by one hundred thousand acres with funding generated by sequestering carbon in forests and are beginning to conduct prescribed burns to protect those forests from megafires. This approach has been shown to be highly effective. More carbon is stored in the world's soil than in vegetation and atmosphere combined; a recent study in *Nature Geoscience* found that prescribed fire not only mitigates wildfires but can stabilize and *increase* soil carbon.

Some pundits claim that climate devastation and destruction are inevitable. We should stop pretending, they argue, that there is any way to prevent a dystopic future of massive crop failures, megafires, imploding economies, biblical flooding, and hundreds of millions of refugees. But many Indigenous communities have already lived through apocalypse: the Yurok described the arrival of white people in 1850 as "the end of the world." Kyle Powys Whyte, a professor of Native American studies and philosophy and an enrolled member of the Citizen Potawatomi Nation, points out that Indigenous people "already inhabit what our ancestors would have likely characterized as a dystopian future." In Whyte's view, climate destabilization is merely the newest assault of settler colonialism, which has inflicted environmental change on Indigenous people for generations. Living through these horrors, Indigenous people have adapted and resisted and organized. They have persisted in upholding the responsibilities and relationships they had to their land. Indigenous people, Whyte argues, know what it means to survive and flourish.

ONE DAY IN Orleans, I sat at a table with Clarence Hostler, a Native elder and high dance ceremonialist, eating turkey sandwiches. We talked about traditional burning. He did not identify as Hupa, Karuk, or Yurok. Those were labels, he told me, created by the federal government. Instead, he identified himself by the village he was born in and the bloodlines he descended from, which extended from Weitchpec up the Trinity River, as well as up the Klamath River to Karuk territory. His great great-uncles had been born before the area was colonized, and he had learned his culture from them. "We learned the old songs young," he said. But it was his grandmother who taught him how to burn. Around the age of three she began bringing him up into the mountains to gather basket materials. At the age of five or six, with his brothers and cousins, he began to help her set fires to bear grass. "Her gathering was high above Weitchpex, along the ridgetop between Kewet, or Burrill Peak, and Onion

Mountain," he said. I had seen such a place, driving in the dark one night down a dirt track, staring up at the shadows of the towering trees in awe. "You can tell it's a place that was burned and managed for thousands of years. It looks different up there," he said. His grandma made bundles out of wild oat grass gathered along the river that were tied with willow bark strips. "My grandma would sing her song over the fire and then give us the bundles and we would go out," he said. "I thought it was forty acres. When I go back now, I realize it's about three." I laughed and asked how often she returned to burn. "Every three years," he said. "She'd harvest the bear grass that first spring and then let it grow." The last time he burned bear grass with his grandma was when he was twelve years old.

Now he imparted advice to me. "Western scientists don't like to say 'people,' but animals and plants were the first people. Humans are the newest people," he said. "When you're out there burning, be mindful that you're helping the people. And talk to them, like my grandma did."

Something I loved about fire was the way it allowed me to time-travel, my perspective telescoping back and forth as I imagined long ago pasts and vivid potential futures. In Orleans, I watched the crew crouched around a big paper map that lay on the ground. It showed a piece of forest scarred by history. On one side it was bordered by the unfinished GO-road that had been built for logging. Massive Gold Rush–era ditches sliced through it. Douglas firs now choked it. Could those gold miners have ever imagined this group of people a hundred years in the future, trying to repair what they had broken? Who would do this work a hundred years from now? Would they think back to us?

On my last day on the Yurok Reservation, we burned a tiny patch of land, just two acres in size, that lay around the home of a local family. The house was made from cedar and sat on top of stilts. Raven explained to me how vulnerable it was to wildfire: the stilts created a space that would suck in the hot air and flames and circulate them. We prepped for our burn by installing sprinklers under the house to increase the humidity in case of any spot fires, and pulled apart "jackpots" of dry fuel in the surrounding woods to try and temper the intensity of the burn.

At the test fire, Margo lit the wormwood torch and sang a song of prayer. Then the burn boss, a man from CAL Fire, gave a prayer thanking Jesus. Lastly, the crew's safety officer, a military veteran with the BIA, offered up a song of prayer in Shoshone. We worked slowly, mindful of the house, not wanting to take any risks. In the afternoon, I took a drip torch and climbed up a bank into the woods to finish the last half acre. The temperature was in the high eighties, and the slope was south facing; it had been warming and drying in the direct sun for hours. We were igniting an upslope head fire into the black; it was going to burn fast and hot. Another ignitor hiked ahead of me, and together we began to lay down small dots of fire in a synchronized pattern. We worked quickly to avoid our own flames, crashing through woody shrubs. The leaves snarled and exploded from even a little dribble of fuel. As the fire built, I began to feel like I was working in an oven. My body poured sweat trying to cool itself. I heaved through snarls of blackberry that cut the skin under my clothing, but I barely noticed. When we reached the handline at the end of the unit, I jumped down the steep cutbank onto the road. I stood and drank water when waves of dizziness and nausea began to hit me. I had known a journalist who died of severe heatstroke, and now I felt a tinge of fear. Where was this going? I swallowed my pride and sat down on the ground, taking off my helmet and pulling down my yellow Nomex around my waist. My face was so crimson that one of the burn bosses came over. "You can take your boots off to cool down, you know," she said, and stood nearby, watching. The dizziness eventually went away. I stood and went to mop up. When we finished, I squished my pack and tool into the back of a little Subaru and climbed in with some of the crew. We drove down a narrow road to the banks of a wide creek. Rocks had been piled in small dams, creating pools of clear water, the surface of which was now covered by a lace of golden leaves. We peeled off our sooty, sweat-soaked clothes and sank beneath the icy water. I put my head under again and again, and the shock and pleasure of it made me gasp. Then I stood on the rocks, shivering. A sense of euphoria overcame me. My body felt so light it was like I was floating above the ground. My body was strong, I realized. It kept going. It was a tool for participating in the becoming of the land. *Sympoiesis*: making-with. Other than when I have given birth, I had never felt so integral to a life-giving process as I did lighting a fire.

We had a feast of salmon that night, the fish cooked in a traditional way over an open pit of embers. Frankie Myers, the vice chair of the Yurok Tribal Council, addressed the crew. "When the first anthropologist came to Yurok territory, there's a few things that they noted: the Yurok people were people in constant prayer and constant fire. As we sit and we gather and we eat, this is a prayer that we're making together," he said. "Once again, thank you for helping us make our medicine again for our landscapes and for us as a people as well. Thank you for your prayers, thank you for coming and praying with us in this way that we do."

I thought back to the beginning of our burn that day, when we'd lit the test fire. When I'd looked up at the house on stilts, I saw two little boys on the porch, wiggling and laughing with excitement as the smoke began to rise. They weren't afraid. Margo saw them too. "They'll have no memory of a time when we didn't burn," she said.

EPILOGUE

A YEAR AFTER I LEFT NEBRASKA, I RETURNED TO THE SAME PRAIRIE TO WORK ON the crew. I flew into Denver again, but this time, instead of heading east toward the town of Ord, I went north, into South Dakota's Black Hills along dirt roads past grazing buffalo. As the sun set, I drove through Wind Cave, the place where the Sioux believe they originated in an underground world. When I finally saw the lights of Rapid City, I found a truck stop off the interstate, paid to take a hot shower, and slept in the back of my rental SUV.

For the next three days, I was enrolled in a wildland-firefighting course in operating chainsaws. On the spectrum of tree hugger to logger, I was full-on hugger. I had never wanted to use a chainsaw. I secretly hated being around them on the fireline, where I could never stop compulsively imagining every potential sliced artery. I admired the skill and stamina required to be a sawyer, but I also suspected they were adrenaline fiends. I was there because of Mike West. Months earlier I had mentioned that I hoped to work on a handcrew again. "Nice!" he said. "Go be their lead saw!" The fact that anyone, let alone Mike, thought I was capable of being a sawyer surprised me. But his confidence shifted the limits of what

I considered possible for myself. Could I really learn to operate a saw?

My understanding of forest management had also evolved—radically. Walking through the forests of California, New Mexico, Arizona, and other places had convinced me of the usefulness of the chainsaw for restoring the resiliency of forests. My sentimentality about killing trees began to feel like an untenable ethical position. So there I was, in a classroom at Black Hills State University, learning about sloping and boring cuts, tension and compression, gunning sights and dogs. My instructors were Eric and Justin, the two best friends I had traveled to the Grand Canyon with the previous fall.

"We've all cut wood with old pappy. We have our own way of doing things," Eric told us on our first day. "As long as it's not wrong, it's right." According to Eric and Justin, the key to being a good sawyer was "trigger time": operating a saw at every possible opportunity, every week, every day, if possible. They taught us how to size up and read trees, how to do complexity analyses and to visualize felling plans. We learned a new vocabulary: pickles, pistol butts, bendies, punky knots, and goofy parallaxes. We covered safe zones and escape zones and blood bubbles. "We're not loggers; we don't give a shit about board feet," said Justin. "We just don't want to die."

At night, I crossed the highway near Black Hills State University and parked at the truck stop, reading under a blanket with my headlamp on so no one could see me from the outside. I had been sleeping in tents and cars so much over the previous year that I now slept better in them than I did in my bed at home. My sleeping bag had become my safe place. Still, I slept with my Leatherman knife out, putting it in the cupholder near my head.

On our third day of class, we drove to the outskirts of Rapid City. It was a twinkling, sunny day. Western bluebirds watched us

Epilogue

from a power line as we unloaded saws from the backs of trucks. The land was crowded with ponderosa pines, most of them young blackjacks. My palms were already sweating with nervousness as I clipped a pair of Justin's green chainsaw chaps around my waist and legs. Everyone spread out and started up the saws. I stood next to Eric, not sure where to begin. To pass the course, we had to cut at least three trees. I thought back to Alfredo Padillo, the priest in New Mexico. "Sometimes caring for God's creations means making choices," he had said. My eye was caught by a tree nearby. It was a multistem split: a smaller trunk was growing up and out of the main trunk at its base. Its crown was actually three separate diverging stems, making its lean slightly difficult to discern. Its diameter was a bit bigger than that of the other trees. I knew right away I would be pissed if someone else cut it down. Fuck it, I thought and turned to Eric. "What about that one?" We walked over and studied it up close. I'd have to cut the smaller trunk growing out of its side to get to the main trunk, which was about sixteen inches in diameter. And before that, I'd have to cut another tree that was in the path of where I wanted this one to fall. It was a three-for-one.

"Do it," said Eric.

I got to work, making face cuts, using my gunning sights, yelling "BACK CUT!" I tried to remember Mike's advice. "It's super awkward at first," he had warned me. "It feels so foreign. But it eventually just becomes like an extension of your body. Just remember to breathe when you're cutting a tree. It's real easy to tense up. Try to be as loose as possible, and let the saw do all the work. You'll be good to go. It'll be a blast.... It's making me want to fire up the saw right now." I tried to keep my arms relaxed, feeling the pull and resistance of the wood in the teeth of the saw. One by one, I felled each stem, and when it was over, three trunks lay on the ground exactly where I had planned. I stood over the stump with its level cuts and good hinge wood, adrenaline coursing, grinning and looking around to catch someone's eye. Did anyone see how great that was?

I cut, bucked, and stumped more trees. I thought I was in good shape, but by the end of the day my arms were weak from the effort. I hadn't "let the saw do all the work." Justin walked over to me. "Do me a favor," he said. "Turn on the throttle as high as it will go and let it run." I started it up and pulled the throttle trigger all the way in, trying not to grimace. After a minute, Justin motioned for me to stop. "The saws *need* exercise," he said. "The saws *want* to go full tilt. They feed on it!"

I got the message: Stop being scared. Let it rip.

THAT NIGHT, MY car shook and trembled as westerly gusts of wind slammed into its sides. Coated in sawdust grime and gas fumes, I slept past sunrise and wandered around a Target to buy some clean shirts. The drive south took me through Pine Ridge Reservation, and I pulled into a dirt road to walk to the cemetery at Wounded Knee. Nearby, a big metal plaque hung from a metal frame, explaining how, in the months previous to the massacre by government soldiers, the Oglala Lakota warriors had been ghost dancing, a practice based on prophecies that it would "bring back the old days of the big buffalo herds." I walked through the gravestones at the cemetery. It was enclosed by a chain-link fence on which hundreds of medicine bundles, their bright colors faded by sun, blew in the wind.

Across the state line in Nebraska, gales began to cut across the highway so hard they tore the plastic splash shield from the car's undercarriage, and I had to pull over and rip the rest of it off. The closer I got to the Loess Hills, the more eastern red cedars I saw. Thick stands had cast their berries into pasture, and they were now

Epilogue

growing into young saplings—a visual progression of the forest that would be there in twenty years' time. In Ord, I parked in front of the familiar storefront on Main Street and walked through the door into the incident command post. Everyone was there: Dan, Patty, Ben, the crew boss trainee, and Zeke. Will Harling from the Klamath, too. The crew numbered almost thirty people, including a contingent of firefighters from Spain and Ecuador. Everyone was excited to burn. But there was a vein of tension among the overhead.

"How are the units?" I asked Dan.

"Disconnected," he said. "Not prepped well."

Earlier in the year, the governor of Nebraska had issued an executive order directing the state's agencies to prevent the implementation of President Biden's mandate to conserve 30 percent of America's lands and waters by 2030. In Nebraska, where 97 percent of land was privately owned, the mandate stoked fears of federal overreach and land takeovers. The governor was traveling to dozens of counties to warn people about the federal program, claiming it would devastate the state's economy. Ben couldn't be sure, but he thought he'd noticed a temporary drop-off in landowner interest in participating with nonprofits. As a result, it had taken longer to find units for the crew to work on, and the ones he did find were an hour away in a different county.

The even bigger problem was the wind. It was still blowing at twenty-five miles per hour with gusts of forty to fifty. Oh, and the drought. Nebraska had gotten little to no snow over the entire course of the winter. The town of Ord was in the dead center of a big red blob on the US Drought Monitor map. Red meant "extreme drought," and in the past, similar conditions had led to fish kills, hay scarcity, sell-offs or even culling of cattle, and abandoned horses. When we went to scout the units, the soil was parched and there were no signs of green-up at the stumps of the grass. The recovery for overnight relative humidity was only around 30 percent. The prairie was ready to explode.

On our first morning of operations, we filled up on fuel, prepped our engines, and started to convoy out to the unit to stage, putting fuel at drop points and locating sites for portable pumps. A few minutes into the drive, Ben got on the radio. Everyone needed to turn around. When we pulled up to the town's square, the fire marshal's red truck was now parked in front of the incident command post. We waited outside for an hour, and then Ben delivered the news: burn permits had been suspended—indefinitely.

In the ten years that crews had been gathering to burn prairie near the Loup River, 2022 would prove the most difficult. The crew burned 90 percent fewer acres than in previous years. And although I didn't know it at the time, it was a sign of a difficult and complicated year for prescribed fire all across America. Many of the places and people I had encountered in the previous year became the center of tragedy or controversy. New Mexico experienced the largest wildfire in the state's history when a burn pile that had been lit months before in the Santa Fe National Forest reignited. Just two weeks later, a prescribed burn in the same national forest began to spot as unexpected and erratic winds blew through and crews lost control. Within days, a catastrophic wind event pushed the two fires together to form what became the Hermits Peak/Calf Canyon Fire, which burned nearly 350,000 acres and hundreds of homes. Hermits Peak was seven times bigger than the Cerro Grande Fire of 2000.

The Forest Service responded to the New Mexico disaster by putting a ninety-day moratorium on prescribed burns on all national forests until a review could be conducted. When the report was published, investigators found that, although the burn plan had been followed, the condition of the forest itself had led to the catastrophe: historic drought, no rain, little snowpack, accumulations

Epilogue

of fine and dead fuels. There had been a narrow window for crew availability, good smoke dispersion, and the right burn prescription, and enormous pressure to make up for years of delays in reducing the fire deficit. "Fires are outpacing our models," said the Forest Service chief. "We need to better understand how megadrought and climate change are affecting our actions on the ground." Legislation had been introduced in New Mexico to ban all prescribed burns in the spring.

My crew boss on the Dixie Fire was sent back to prison. Then the private contracting company I had worked with in California lost a crewmember. He was twenty-six years old, a single dad with a young son, and had been a wildland firefighter for five years. While he was working to prepare an area for a prescribed burn, a dead tree fell and killed him. The company's crews escorted his body in a convoy two hundred miles home, while first responders stood along the highways and overpasses to pay their respects.

Thom Taylor had told me that there was always going to be inherent risk in a job that required spending so much time in the forest. "If you work in the woods, shit's going to take you out," he said. When he told me this, he was about to go into his twenty-ninth fire season. He works as a fire operations specialist in the Payette National Forest, often on night operations. His job requires orchestrating resources and crews while putting a lot of fire on the ground. But working at night, away from the fire's edge, helps him feel in control, like he can manage risk better. I had always wondered why, though, after everything he had experienced, Thom stayed in fire. One day, I finally asked him. There were a few different reasons, he told me. He sees himself as essentially a ski bum with a cool summer job and "fire and all that shit is super addictive." He also loves to teach and has made it his mission to try to positively shift the culture of wildland firefighting through training and leadership out on the fireline. But, when it came down to it, there was no way he could quit.

He feels a responsiblity, sometimes to a debilitating degree, to be a voice for the fallen, the ones on the rock scree next to him who did not make it out of their shelters. "Thirtymile is a part of me," he said, "And it's a part of who I am to make sure Tom, Karen, Devon, and Jessica have a voice. As my mom would say: 'If not you, then who?'"

Soon after the young firefighter's death, a fire that was suspected to be human caused started on the eastern side of Mariposa Grove in Yosemite, near the same trail and dense, overgrown forest that Steve Pyne and I had hiked through less than a year before. Sprinkler systems were installed around Grizzly Giant in case the fire made a run into the heart of the "solemn temple."

For a couple of weeks, the national news media covered the Yosemite story with headlines about firefighters in a pitched battle to save the ancient trees. But the Washburn Fire never entered the Mariposa Grove; it burned nearly five thousand acres in a progression that seemed to prove in real time the critical necessity for better land management through fire and forest thinning. A map compiled by Kent van Wagtendonk, the son of the fire ecologist Jan van Wagtendonk, who had pioneered Yosemite's "prescribed natural fire" program, showed the Washburn Fire literally snaking around land that had received controlled burns or fuels treatment, or had undergone a wildfire in the last twenty-five years. Then, on July 16, Jan van Wagtendonk passed away from cancer. The fact that his decades-long effort to bring fire back to the park was saving the Mariposa Grove in the days before he died was described as "poetry" by his son. "Ironically, the Washburn Fire burned Jan's plots just a few days ago," said Dan Buckley in an email. "Kent and I believe he was out there the other night with the firefighters with his hands on the torch."

But the positive outcome of the Washburn Fire was not covered by the media nearly as much as its initial threat to the giant

Epilogue

sequoias. Nor was it widely reported that a month before the fire ignited, an environmental group had filed a lawsuit against the park to halt the very kind of fuels treatment that had helped to protect the Mariposa Grove from devastation. The group's director told a reporter that the park was clear-cutting mature and old forests in Yosemite Valley as part of a commercial logging program. Pyne told me he thought the timing of the lawsuit and the Washburn Fire was "creepy." I agreed. The Washburn Fire and so many others became apocalypse clickbait, sowing fear and dread that could be leveraged by ideologues while the complex but incredibly hopeful story—the one about ecology, Indigenous knowledge and survival, and the good work of folks like Jan van Wagtendonk and Dan Buckley—felt like it slipped past another news cycle.

AFTER OUR BURN permits were suspended in Ord, we spent the following days listening to presentations, practicing fire shelter deployments in the wind, tinkering with equipment, and waiting for a change in weather. I suspected everyone was praying for rain each night. I was. Ben disappeared from the incident command post to pore through burn plans and call landowners from his office, hunting for opportunities to put us to work. Finally, on the sixth day, he found one. It was in a nearby county where the fire chief was willing to issue us a one-day permit to burn a 210-acre unit owned by a local rancher. We went to look at it. It was a stunning piece of land. The rancher had rested it for a year and hadn't allowed any grazing on it. As a result, the hills were covered in cured grass so lovely and thick it looked and felt like a ragged mane. On the day of the burn, we parked our vehicles in a wide dirt area, engines and pumps running, for morning briefing. We knew there was a lot at stake. The community's trust. Ben's and Dan's reputations. The landowner's investment. "I've got full confidence in you," Dan told us. "Just do the next right thing. Right? Be deliberate. Be cautious. Communicate. Do the next right thing."

We started in the southwest corner of the unit with two task forces moving in opposite directions. I was the first ignitor on Patty's task force and brought fire with me along a barbed-wire fence. Two ignitors followed behind, laying fire in strips staggered ten feet apart. I turned and watched the smoke begin to billow and grow into a white mass over our heads. When I reached the southeast corner, we laid fire to a patch of reeds and watched the thirty-foot flames and sparks erupt like a volcano spitting magma. Then I handed my torch to another ignitor and started hiking north into the middle of the unit, alone. I carried my fire pack, radio, hand tool, notebook, and pencil. In the far distance, I could see Will Harling strolling along the bottom of a giant hill, lighting a head fire up its slope that sent up a curtain of silky black smoke. I kept parallel with him and watched as his backing flames and the undulating heat they generated moved my way.

The fire and I met near the banks of a small lake. Approaching a free-roaming fire on the prairie felt just like meeting a horse for the first time. I walked toward it without fear but fully alert, showing myself to the fire as it showed itself to me. Then I stood at the lip of the wave of flame, watching as it consumed the grass—hungry, feral, and free. I sat down on the earth and listened to it burn. I scribbled things on paper: rate of flame spread, smoke observations, fire behavior, wildflower names. I gathered and bunched dried stalks of grass and lit the bundle on fire to make a torch, feeding the flames and spreading them around, loose-herding the fire. I had a funny feeling, like the fire and I were friends and we might just spend the day together, hanging out. Ben appeared at the top of the hill and walked toward me. We laughed at one another in recognition. He wanted to hang out with the fire too. "Sometimes I like to come

Epilogue

out here and just turn my radio off," he said. "I feel like I could fall asleep listening to this sound. It's like the crackle on an old record player."

We walked east through the grass for a while. I saw my task force lighting an enormous flanking fire coming toward us, and we shifted our course to make sure we were one step ahead of its path. In places, the flames reached eight feet, frolicking and galloping across the tall grass, not a horse but a whole rag of spunky colts. We saw Dan's truck and Patty's UTV parked on a hill to the north. At the top, Patty handed me a Very pistol. "Firing!" I yelled, and pulled the trigger. The sparks shot through the air like a comet. I set off again to find my task force. There were two women from Spain, a mustachioed fire chief from Kansas, and two kids. Patty joined us. Ben's six-year-old son held the hose on her UTV and laid down a wet line with help from a thirteen-year-old local kid. Behind them, the fire chief laid down a blazing strip of fire. We chatted and laughed and helped the kids. I had a vision of a band of people walking across a burning savannah ten thousand years ago. How odd, I thought. None of us were related by blood or came from the same place. Our only connection was the fire we loved to light together.

I heard Jason Nez, the archeologist at the Grand Canyon, tell a story once about how it was through a fire in the night that scattered clans of people were able to find each other and become the Diné. "We were wandering around the whole Southwest, and off in the distance we saw a fire on the horizon and started going there," said Nez. "Every day, we'd stop and camp and sit in the dark, and we'd look for where that fire was and head toward it. That's how our clans found each other. We changed from being a scattered people, groups of tribes, and we came together as one big nation. That's what fire can do. It can guide us; it can bring us together. Is it nature? Is it chance? Is it divine intervention? I don't know. That's what we do in fire, is we come together."

Nez believes that the next generation will fix our ailing landscapes. When he trains young people in fire, he tells them, "As we work in fire, don't be afraid to let yourself shine. We're all artists. We're all storytellers. We're all painters. We're all dancers. Our first thought of fire—it's not fear. On deep levels, we know how to use fire, we know how to work it, but we just keep it way down because we haven't had a chance. Our first thought of fire is usually awe."

In the afternoon, the entire crew gathered at the northeast corner of the prairie as the two task forces' fires moved toward one another and then converged like a seam zipping closed. Dan stood between the two flame fronts like Prometheus himself, hands outstretched for warmth. I walked back through the middle of the unit, up and over the smoldering hills, picking up scorched cattle bones and stopping to look at the world we had transformed into burnt lands. Fire transmutes all it comes into contact with, even people. Fire had changed me. I had undergone an aesthetic revolution since driving in Australia years earlier. Here was a beautiful moonscape that would soon proliferate in a kinship between fragrant grasses and wildflowers, insects and birds, a rhythm of disturbance and loss, promise and causality from which I was not separate.

Dan drove out to the unit before dawn the next day to check on our work. When he got back to the incident command post, he told us, "The sun was just peeking through. I just wanted to drop my tailgate and make some coffee with that million-dollar view. It looks wonderful."

MORE DAYS OF waiting followed. With few options and no rain predicted, we focused on blacklining. Even if we couldn't broadcast burn ourselves, blacklining would give the landowners more opportunity and security to set fires themselves later on. "We all want to burn big. We all want to see big fire," said Dan. "Get creative and get some shit done." I found that I was less entranced by the drip torch than

Epilogue

I had been the year before. Instead, I commandeered a UTV and focused on learning every single thing about its pump: how to start it, fill the tank, fix a hose leak, winterize it each night. I made dumb mistakes. I lost my fire pack when it fell out in the prairie; a spare tire rolled down a hill I was driving up; I got the UTV's back tires stuck in a giant gopher hole and Dan had to pull me out with his truck. Patty loved my gaffes. "At least it's not me!" she said gleefully. Fire also continued to reveal new strengths. I held the backdoor for the crew, whizzing around, steering with one hand and holding a nozzle in the other, tool always at the ready, stopping, chopping, digging, stirring, stamping, spraying, speeding, backward, forward, putting out smokes and flames for hours. I worked at full song.

Sometimes I drove with Dan on long errands across the county to scout units or pick up equipment. We ventured into a windbreak of mature cedar trees forty feet thick just to see what it was like inside, crawling on our hands and knees under scraggly branches and on denuded ground where nothing could grow. Even in the dark interior of the trees, the wind wailed.

Our conversation often meandered. He was always trying to think of creative ways to solve the fuel-loading problem. He would think out loud: What if we could move portable power generators around prairies and forests that could burn the biomass and generate electricity? Then he would sigh. "Sometimes I wonder if we missed our chance. I worry we're too late."

Whenever I could, I stayed close to Patty. She was a fount of know-how, blunt honesty, and jokes. She seemed to tolerate me hanging around in her shadow. One freezing morning, we set up a portable pump at the edge of a muddy pond. I filled a plastic jug with icy water to prime the machine again and again. Crouching next to each other in the mud, I noticed how the back and shoulders of Patty's fire shirt were patched and held together with white stitching from the zigzag pattern on a sewing machine. Her hands were red with the cold and bandaged with band aids and strips of the same duct tape that held her radio together. We struggled for a bit, and finally the pump grumbled to life. Patty was in disbelief that I could do anything useful. "She's grown a lot since we got her last year," she told the overhead later, when I was in earshot. "She was actually helping me set up the pump!" She turned and shouted at me. "You're not a *journalist!* You're a wildland firefighter!"

The next morning, I arrived at the command post in the dark. Patty was there, alone. She looked tired. I asked if she was OK. She hadn't slept much, she said. She'd gotten a phone call late the night before telling her that a close friend back home had been killed by a falling tree. I could see the pain on her face. It didn't seem to occur to Patty that she could take the day off. If anything, she was more indomitable, racing everywhere on her UTV, even grabbing a tool and digging line alongside us, like work was the key to salvation.

Out on the unit that afternoon, a weather report came in over the radio. The relative humidity was rising. Patty looked at the sky. "I'm thinking thunderstorms," she said. Sure enough, over the next hour, dark clouds appeared in the horizon and moved toward us, pregnant and quivering with lightning. We needed to get back to the trucks and a paved road before the dirt tracks became slick and impassable with mud. "You want to ride with me?" Patty asked. I hopped onto the seat of the UTV next to her and held on as she shot across the yellow hills. Above us the sky had become a silver dome. It began to roil. We felt the first, fat drops, and then thunder erupted. The wind whipped our hair. An icy hail began to hammer the ground and our hardhats. I whooped at the storm. Here was the rain we had prayed for. Everything that had seemed impossible had become possible. Our window for fire was opening. *The earth, the earth that is coming*, I thought. *Wonderful the earth that is coming.*

ACKNOWLEDGMENTS

In the course of reporting this book I crossed paths with many women and men on the fireline who generously gave me instruction and demonstrated an incredible commitment to the land. Thank you to those both named in this book and unnamed for allowing me to learn by your side.

My great thanks to Michelle Tessler, a champion who has continuously helped me to hone and improve my ideas for more than a decade. Thank you to Remy Cawley for sharing a vision and supporting this book early on. I am indebted to Lisa Kaufman, an unparalleled editor: thank you for your *nachas*. And much gratitude to the team at Bold Type Books and Hachette for shepherding the book to publication, especially Kaitlin Carruthers-Busser, Anupama Roy-Chaudhury, and Kelley Blewster.

I am profoundly grateful to Steve Pyne for letting me tag along and ask so many questions over the years; this book would not have been possible without your groundbreaking scholarship and perspective. Several other scholars and researchers who I am thankful for are Tom Swetnam, Nate Stephenson, Christy Brigham, Michael Fromm, Scott Stephenson, Susan Pritchard, John Ortmann, and

Crystal Kolden. Lastly, Frank Lake for sharing his incredible insights and work. Thank you to Lenya Quinn Davidson for her early guidance and a fortuitous connection to the wonderful Heather Heward at the University of Idaho. Sincere thanks to Jeremy Bailey for helping me to navigate task books and quals and reassuring me at a moment of doubt that there is a need for both "burn bosses and storytellers." Big thanks to Ben Wheeler for his infectious love of the prairie and his friendship—I can't wait to go on a roll together. And my appreciation for Melanie Boden, crew mother, for her casseroles, birthday cakes, grape salad, and hugs.

I am indebted to Dan Kelleher and Patty Carrick for giving me the best life goals; thanks for being my teachers. Many thanks to the folks I worked with on crews, and especially Jeff Priebe, Neil Chapman, Ethan Huebner, Augustine Beard, Martin Jacob, Ryan Lodge, Eric Barnum, Miranda Flora, Jason Barker, Dylan Bryan, Brian Faith, Anna Higgins, Chris Root, José Luis Duce, Mike Beasley, Will Bruce, Thea Maria Carleson, Andrea Mariá, Ghaleb Attrache, Jane Manning, Jared Childress, Clare Lacy, Brian Peterson, Zeke Lunder, Will Harling, Kelly Martin, Brian Anderson, Eytan Krasilovsky, Leonora Pepper, Sam Berry, Michelle Ly, Annabelle Tao, Bob Bale, Mike Rozdolksi, and Raven Parking.

Special thanks to: Tyler Briggs for being the nicest burn boss in fire. Eric O'Connor for showing me Rapid City's finest middle school orchestras and Justin Bauer for letting me wear his chaps: you guys are the best. Claire Brown and Max Brotman for bringing joy and hilarity to the fireline. Bre Orcasitas for her strength and friendship. Mike West for his vulnerability and generosity; looking forward to that orange LaCroix. Josh Heller for bringing me on board and the whole of Crew Six for showing me how it's done. Dan Buckley for sharing the benefit of his decades in fire. Margo Robbins and Elizabeth Azuzz for demonstrating a love of fire. Thom

Acknowledgments

Taylor for his incredible honesty and willingness to share his story. Jason Nez for his passion for the past and hope for the future.

Also, thank you to: Emma Piper-Burkett for giving me a place to rest in Boulder and the gift of her friendship. Mo Parker for being a fantastic aunt. Janet and Chris Miller for their love and unconditional support, which has made all the difference at critical moments. Mark Miller for his conversation and rides in his Mustang to the airport. Thanks to my dad, Rory O'Connor, for his enthusiasm and support. My mom and stepdad, Katherine Miller and Robert Heinzman, for taking trains, planes, and automobiles to be wonderful grandparents and cheerleaders.

Lastly, thank you to Joaquín and Tareq for going on adventures, making me laugh, and giving me a beautiful purpose in life. And, finally, love and gratitude to Bryan Parker, the greatest partner, who makes everything possible.

NOTES

PROLOGUE

2 **Gammage documents the complex:** Gammage, *The Biggest Estate on Earth*, 122, 131.
3 **"super outbreak":** Peterson et al., "Australia's Black Summer Pyrocumulonimbus Super Outbreak."
3 **"Many of us":** Maguire, "Bushfire Destroys Couple's North Rosedale Home."
4 **"If people don't take":** Lewis, "Ecological and Technological Knowledge of Fire."
7 **"Nature, in this conception":** Nussbaum, "A Peopled Wilderness."
7 **"essential element for ongoing":** Haraway, *Staying with the Trouble*, 44.
7 **"In some places where the Indians":** Stewart, *Forgotten Fires*, 76.
7 **"America was not a virgin":** Pyne, *Between Two Fires*, 53.
8 **by as much as 50 percent:** Bowman et al., "Human Exposure and Sensitivity to Globally Extreme Wildfire Events."
10 **"Fire is a gift from Creator":** Stephens, "Fire on the Mountain (and Elsewhere)."
10 **states intentionally burning:** Geographic Area Coordination Centers, "National Year-to-Date Report on Fires and Acres Burned."

ONE

14 **"Fight fire aggressively":** National Wildfire Coordinating Group, "10 Standard Firefighting Orders."

14 **"If you're under the impression":** National Wildfire Coordinating Group, "Module 1: Basic Concepts of Wildland Fire."
18 **"You sing the country":** Gammage, "Fire in 1788."
34 **"Then as the things of the earth":** Murie and Parks, *Ceremonies of the Pawnee, Part I*, 53, 47.

TWO

44 **"greening green":** Marder, *Green Mass*, 33.
44 **"O most honored Greening":** Williams, "The Liturgy of Home."
45 **"The sky was split in two":** Phillips, "The Tunguska Impact."
46 **"There could have been no":** Goldring, "The Oldest Known Petrified Forest."
48 **"evolutionary arena for social":** MacDonald et al., "Middle Pleistocene Fire Use."
48 **"tactile and real":** Chazan, "Toward a Long Prehistory of Fire."
50 **"sequoias" of the east:** Dyer, "Recoding the Chestnut."
51 **Kudish thinks it is possible that some 95 percent:** Trapani, "First Growth Forest in the Catskills."
51 **oak bottleneck:** Alexander et al., "Mesophication of Oak Landscapes."
52 **"People want to believe":** Deloria, *Red Earth, White Lies*, 68.
53 **Stewart became a participant-observer:** Howell, *Cannibalism Is an Acquired Taste*, 84, 34, 148.
54 **historical accounts Stewart gathered:** Stewart, *Forgotten Fires*, 82, 75, 80.
55 **"thunder rumbling as it goes":** Catlin, *Prairie Meadows Burning*.
55 **"The prairies burning form":** Stewart, *Forgotten Fires*, 116.
56 **"The extensive grasslands":** *Ibid.*, 69.
56 **"face up to the removal":** Anderson, *Tending the Wild*, 41.
57 **"An abundance of evidence":** Stewart, *Forgotten Fires*, 149.
59 **ninety-nine fire scars dating:** Hoffman et al., "Ecological Legacies of Anthropogenic Burning."
59 **Scientists call landscapes:** Sullivan and Forste, "Fire-Reliant Subsistence Economies."
60 **"course of an Indian":** Lewis and Ferguson, "Yards, Corridors, and Mosaics."
60 **thirteen million acres once burned:** Anderson, *Tending the Wild*, 136, 239.
61 **"pruned orchard":** Tripp, "Cultural Resource Message."
61 **"Fire is life":** Azuzz, "Rekindling Old Relationship."

THREE

65 **"sweet perfume":** Collins and Anderson, *Plant Communities of New Jersey*, 55.
67 **Butterfly Station:** Barnes, *Natural History of the Albany Pine Bush*, 41.

Notes

67 **"Wide open on its pin"**: Boyd, *Vladimir Nabokov*, 53.
68 **"ought to be out"**: Mittelbach and Crewdson, "Chasing Nabokov's Elusive and Endangered True Love."
68 **"A score of small"**: Nabokov, *Pnin*, 128.
69 **A Dutch colonist described how**: Barnes, *Natural History of the Albany Pine Bush*, 21, 22, 3, 24.
70 **"sandy and flowery"**: Mittelbach and Crewdson, "Chasing Nabokov's Elusive and Endangered True Love."
70 **"zone of truck yards"**: Zimmer, "Butterflies and Moths Named by and for Nabokov."
70 **declined by 99 percent**: Albany Pine Bush Preserve Commission, "Endangered Karner Blue Butterfly Population Boom."
75 **Flematti hypothesized that seed plants**: Flematti et al., "What Are Karrikins?"
76 **estimated 46,100 butterflies**: Albany Pine Bush Preserve Commission, *Annual Report*.
76 **"But now I settled"**: Thoreau, "Thoreau's Journal (Part II)."
77 **"I suppose I would"**: National Wildfire Coordinating Group, "2022 WOR Day 7."
77 **"It's setting your ego"**: Viktora, "'Student of Fire.'"
78 **"The line between firefighter"**: Desmond, *On the Fireline*, 61, 63.
78 **"Nature was here something savage"**: Thoreau and Hoagland, *The Maine Woods*, 94.
79 **One day, I picked up**: Lewis and Yarnell, *Pathological Firesetting*, v, vi, 15, 363.
81 **In 1931, two halls of a reformatory**: Zipf, *Bad Girls at Samarcand*, 46, 106, 107.
82–83 **"One of the defendants"**: "Arson Defendants and Their Counsel."
83 **according to the historian Johannes Dillinger**: Dillinger, "Organized Arson as a Political Crime."
83 **Newspapers and official reports were full**: Frierson, *All Russia Is Burning!*, 261, 134.
84 **self-immolations occurred**: Michelsen, "The Political Subject of Self-Immolation."
84 **As many as a hundred South Koreans**: Biggs, "Dying Without Killing."
84 **"My early death by fossil fuel"**: Correal, "'Very Real, and Very Haunting.'"
85 **"To burn oneself by fire"**: Hanh et al., *Vietnam*, 106.

FOUR

88 **"Woods burning had no more place"**: Pyne, *Tending Fire*, 105.
89 **"Bit by bit the old fire regime"**: Pyne, *World Fire*, 17.

89 **"only in this way could"**: Kosek, *Understories*, 160.
89 **"These facts do not imply"**: Agee, "Introduction to Gifford Pinchot's Article."
90 **"Chinamen"**: Kosek, *Understories*, 156.
90 **"not only the underbrush"**: Muir, *The Mountains of California*, 150.
90 **As the sociologist Kari Marie Norgaard described**: Norgaard, *Salmon and Acorns Feed Our People*, 94, 97.
91 **"My scheme is as follows"**: Williams, "Wildland Fire Management in the 20th Century."
91 **These proponents of intentional fire-setting**: Pyne, *Between Two Fires*, 18.
92 **"Nature has always taken care"**: California Fire Science Consortium, "Ogle Defends 1920's 'Light Burning.'"
92 **traveled some three hundred thousand miles**: Rooney, "Burnin' Bill and the Dixie Crusaders."
92 **In one report published in 1940**: Shea, *Getting at the Roots of Man-Caused Forest Fires*, 11, 17, 5, 37, 75.
93 **"ancestor worship"**: Yeater, "Incendiarism," 4.
94 **"an experiment on a continental scale"**: Pyne, *Between Two Fires*, 13.
94 **describes how America's forest landscapes**: Kosek, *Understories*, 197, xiv.
95 **"Cold War on fire"**: Pyne, *Between Two Fires*, 13.
95 **In the 1990s, environmental organizations**: Kosek, *Understories*, 157, 186.
95 **some 40 percent of the club's**: Beirich and Potok, "Countering Anti-Immigration Extremism."
96 **"big-ass fires"**: Anchor Point Media, "Wildfire Fire Saved My Life."
97 **Journalists covering the Cebollita**: Petty, "Help Came Quickly in Jemez Fire."
97 **"I was in Tokyo"**: Carmack, "A Visit to a Tragic Scene."
100 **"When a fire develops"**: McRae, "An Error of Definition."
100 **began calling large, complex, and intense fires "megafires"**: Buckland, "What Is a Megafire?"
104 **"Despite extensive scientific evidence"**: Ingalsbee, "Whither the Paradigm Shift?"
105 **described his studies in a place called the Blodgett Experimental Forest**: Stephens, "Fire on the Mountain (and Elsewhere)."
106 **somewhere between 1 and 5 percent**: Jung, "Not All Large Fires Are Bad."
106 **75 percent tree mortality**: Berger et al., "Fire FAQs."

FIVE

109 **In his book *Young Men and Fire***: Maclean, *Young Men and Fire*, 28, 300, 298, 7, 294.
112 **"the wonders of symbiosis"**: Tsing et al., *Arts of Living on a Damaged Planet*, M2.

Notes

113 **"I remember we spiked":** Anchor Point Media, "PTSD and Why I Had To Leave."
118 **"Humidities will be bottoming":** United States Forest Service, *Nuttall Complex Fire Shelter Deployment Review.*
118 **the lookout later recalled:** WildlandFireLLC, "Nuttall Fire Episode One."
120 **"Cut as fast as you can":** "Nuttall Fire Series."
121 **"Hey, we just got cut":** WildlandFireLLC, "Nuttall Fire Episode One."
122 **"We started bucking stuff":** "Nuttall Fire Series."
124 **"At that point in time":** National Wildfire Coordinating Group, "WFSTAR: Alabaugh Fire Deployment Story."
124 **"It's hard to stay in":** WildlandFireLLC, "Inside the Shelter."
124 **"It's just real scary":** National Wildfire Coordinating Group, "WFSTAR: Holloway Fire Deployment."
125 **"Deploy! Deploy!":** WildlandFireLLC, "30 Mile Fire Fatality Case Study."
126 **Screw that, I'm taking:** "Nuttall Fire Series."
127 **what he described as heart rolls:** Thom, "Hello Darkness...."
127 **"embracing the suck":** Wildland Fire Jobs, "How to Have a Successful Season."
127 **"Everything is earned"** Humphries, "Wildland Fire Leadership."
128 **"A true commitment":** Gabbert, "Experienced Wildland Firefighter Explains Why He Resigned."
131 **"[Nuttall was] the first time":** "Nuttall Fire Series."
131 **"Is LCES Dead?":** Taylor, "Is LCES Dead?"
133 **"In my career, I was almost":** Grassroots Wildland Firefighters, "Mike West Letter."

SIX

141 **"It goes to show":** Nature Conservancy, "Through the Fire."
141 **"The Dragon takes a nap":** Lunder, "Dixie Fire—Friday the 13th."
142 **"I am not used to seeing":** Lunder, "Dixie Fire Severity—8/12/2021."
142 **"Well, things are really":** Lunder, "Dixie Fire—8/5/2021—Afternoon."
142 **"I don't really know":** Lunder, "Dixie Fire—8/4/2021—Evening."
143 **"It's easy to put too much":** Lunder, "Dixie Fire—Firing Operations."
145 **"Perpetual dark sarcasm":** "Montana and California Go Big."

SEVEN

173 **"some of the steepest, wildest ground":** Lunder, "Dixie Fire Severity—8/12/2021."

EIGHT

185 **It is combustion "no longer bounded":** Pyne, *The Pyrocene*, 4, 84, 119, 5.
187 **"A tougher question":** Lunder, "Dixie Fire—Firing Operations."
188 **This "ecology of attention":** Crawford, *The World Beyond Your Head*, 24, 26.
191 **It was, he wrote, "good, tough work":** Pyne, *Fire on the Rim*, 49, 155, 93.
192 **"The other elements":** Pyne, "Commentaries, Otherwise Unpublished."
193 **"has the advantage of showing":** Pyne, *The Pyrocene*, 146.
198 **"damned uphill and down":** Stoddard, "Use of Fire in Pine Forests and Game Lands."
198 **"fire counterculture":** Pyne, *Between Two Fires*, 36, 40.
199 **"biotic associations":** Leopold et al., *Wildlife Management in the National Parks*, 3.
200 **"living in the stupid zone":** Quillen, "Change We Could Believe In."
200 **"By the end of the 1980s":** Pyne, *Between Two Fires*, 228.
200 **a concept of "prescribed natural fires":** van Wagtendonk, "The History and Evolution of Wildland Fire Use."
200 **"wilderness character":** Jenkins et al., "The Evolution of Management Science."
202 **"I can easily imagine":** Bailey, "Stop Focusing on Ignitions."
203 **"sublime wilderness":** Muir, *Our National Parks*, 226.
205 **"Their primary contribution":** Pyne, "Squaring the Triangle," 5.

NINE

207 **80 percent of the state's groves:** Stephenson, "Reference Conditions for Giant Sequoia"
207 **an estimated 10 to 14 percent:** Stephenson, "Preliminary Estimates of Sequoia Mortality."
209 **"unimpaired for the enjoyment":** "Yosemite Logging Unprecedented."
211 **more severely burned groves might benefit:** Meyer and Safford, "Giant Sequoia Regeneration in Groves."
214 **"This long-term view":** Swetnam, "Fire History and Climate Change in Giant Sequoia Groves."
214 **warned in his seminal report:** Leopold et al., *Wildlife Management in the National Parks*, 4, 8.
216 **"reintroducing disturbance":** Anderson, *Tending the Wild*, 350, 354.
217 **"paradigm shift":** Hagmann et al., "Evidence for Widespread Changes."
218 **"I enjoy being in":** Sabalow and Kasler, "Wildfire Experts Escalate Fight."
218 **"The call by non-Indigenous":** Halsey, letter to the editor.
222 **In 1876 there were an estimated:** Covington, "Postsettlement Changes in Natural Fire."

Notes

TEN

231 **"You must tend to the":** Kosek, *Understories*, 122.
233 **the "missing fire":** Haugo et al., "The Missing Fire."
238 **"Outguessing Mother Nature":** Schroeder and Buck, "Fire Weather, PMS 426-1."
242 **"The fire burned large":** Balice et al., "Burn Severities, Fire Intensities, and Impacts."
243 **"one of the worst examples":** Hughes, "It's Rare for a 'Prescribed' Fire to Get Away."
244 **"seting the fir trees":** Journals of the Lewis & Clark Expedition, "Wednesday, June 25, 1806."

ELEVEN

247 **"In Florida fire season":** Pyne, *Florida: A Fire Survey*, 3.
248 **"We allowed the ancient":** Stowe, "The Role of Ritual and Ceremony in Wildlands Conservation."
249 **"peculiar fire heritage":** Pyne, *Fire in America*, 159.
250 **"These trees were doubtless":** Harper, *Economic Botany of Alabama*, 40.
250 **"cut out and get out":** Fickle, *Green Gold*, 63.
250 **"wimmen folks":** Stoddard, "Use of Fire in Pine Forests and Game Lands."
252 a **"happy" place:** Beadel, "Fire Impressions."
253 **"There is sound reason":** Brenner and Wade, "Florida's Revised Prescribed Fire Law."
253 Long **"must have been":** Harper, "Historical Notes on the Relation of Fires to Forests."
254 **"Somebody's been burning":** Way et al., *The Art of Managing Longleaf*, 83.
256 **"pocosin crashing":** Blevins and Schafale, *Wild North Carolina*, 136, 137.
259 **"fire maverick":** Anderson, "Introduction to Omer C. Stewart's Article."
259 **The main problem:** Stewart, "Barriers to Understanding the Influence of Use of Fire."
260 **"The late 1950s and 1960s":** Way et al., *The Art of Managing Longleaf*, 134, 51, 149, 122, 161, 152, 101, 151.
261 **"art of advanced burning":** Beadel, "Fire Impressions."
261 **"folk-wisdom":** Komarek, "The Use of Fire."
262 **"Man has used both":** Komarek, "Fire—and the Ecology of Man."
263 **"That fire stopped right":** Way et al., *The Art of Managing Longleaf*, 132.
263 **"I don't think we can possibly":** Thompson, "Fire and the Environment—Application Section."

TWELVE

- 274 **"doctors of the world"**: Buckley, *Standing Ground*, 215.
- 275 **wrong to think of regalia as "objects"**: *Ibid.*, 175.
- 276 **"The wealthy ones that own"**: Thompson, *To the American Indian*, 116.
- 276 **"world renewal cult"**: Kroeber and Gifford, *World Renewal*, 107.
- 277 **"These are our prayer items"**: Fimrite, "Yurok Indians Exult at Return of Sacred Cache."
- 278 **"government-subsidized capitalism"**: Buckley, *Standing Ground*, 181.
- 279 **the "GO-road"**: *Ibid.*, 48.
- 279 **"We want the sacred country"**: *Ibid.*, 185.
- 280 **no living children were born**: McCovey and Salter, *Medicine Trails*, 205.
- 281 **"almost at the end of the road"**: Anderson, *Tending the Wild*, 112.
- 281 **"When I was a young man"**: Beck, "The Yurok Tribe Is Using California's Carbon Offset Program."
- 282 **"great conflagration that will"**: Thompson, *To the American Indian*, 132.
- 282 **over half a million acres burned**: Haugo et al., "The Missing Fire."
- 283 **"precipitation whiplash"**: Kauffmann and Garwood, *The Klamath Mountains*, 115.
- 283 **average temperature from 1931 to 2014**: Jules et al., "The Effects of a Half Century of Warming."
- 284 **"So they go on like that"**: McCovey and Salter, *Medicine Trails*, 239.
- 285 **"cultural fire practitioner"**: "AB 642: California Assembly Bill, 2021–2022 Regular Season."
- 289 **"against the law"**: Washburn, "Symmetry Analysis of Yurok, Karok, and Hupa Indian Basket Designs."
- 290 **a single Klamath village**: Levy, "Rekindling Native Fires."
- 290 **"When I lived up the Klamath"**: *Ibid.*

THIRTEEN

- 298 **California's burn windows**: Baijnath-Rodino et al., "Historical Seasonal Changes in Prescribed Burn Windows."
- 298 **reduce opportunities for prescribed fire in the Southeast**: Kupfer et al., "Climate Change Projected to Reduce Prescribed Burning Opportunities."
- 298 **windows across the Great Plains**: Yurkonis et al., "Seasonality of Prescribed Fire Weather Windows."
- 305 **"For Native Americans"**: Treuer, "Return the National Parks to the Tribes."
- 305 **Bureau of Indian Affairs was the only federal agency**: Kolden, "We're Not Doing Enough Prescribed Fire."

306 **BIA has allocated millions of dollars:** Bureau of Indian Affairs, "Fiscal Year 2015 Reserved Treaty Rights Lands Plan."
306 **A five-hundred-acre prescribed burn:** Russell et al., "Doing Work on the Land of Our Ancestors."
306 **Eighteen percent of the planet's land:** Conservation International, "Center for Communities and Conservation."
306 **80 percent of its biodiversity:** Sobrevila, *The Role of Indigenous Peoples*.
306 **Seventeen percent of the world's:** Frechette et al., *A Global Baseline of Carbon Storage*.
306 **More carbon is stored:** Pellegrini et al., "Fire Effects on the Persistence of Soil Organic Matter."
307 **"the end of the world":** Buckley, *Standing Ground*, 9.
307 **"already inhabit what":** Whyte, "Our Ancestors' Dystopia Now," 2.

EPILOGUE

317 **"Fires are outpacing":** United States Forest Service, "Statement from Chief Randy Moore."
318 **"solemn temple":** Roosevelt, letter to John Muir.
318 **"poetry":** "Remembering Jan van Wagtendonk."
321 **"We were wandering around":** NPS Western Regions Wildland Fire Resource Advisor Course (N-9042).

BIBLIOGRAPHY

"AB 642: California Assembly Bill, 2021–2022 Regular Season: Wildfires." Open States, accessed March 14, 2023. https://openstates.org/ca/bills/20212022/AB642/.

Abrams, Marc D. "Prescribing Fire in Eastern Oak Forests: Is Time Running Out?" *Northern Journal of Applied Forestry* 22, no. 3 (September 1, 2005): 190–196. https://doi.org/10.1093/njaf/22.3.190.

Agee, James K. "Introduction to Gifford Pinchot's Article." *Fire Ecology* 7, no. 3 (December 2011): 2. https://doi.org/10.4996/fireecology.0703002.

Albany Pine Bush Preserve Commission. *Annual Report*. April 1, 2020. https://albanypinebush.org/fckimages/uploads/1623362147_APBPC%20annual%20report%202021.pdf.

———. "Endangered Karner Blue Butterfly Population Boom." July 14, 2020. https://albanypinebush.org/connect-press-room&prrid=53.

Alexander, Heather D., and Mary A. Arthur. "Implications of a Predicted Shift from Upland Oaks to Red Maple on Forest Hydrology and Nutrient Availability." *Canadian Journal of Forest Research*, April 15, 2010. https://doi.org/10.1139/X10-029.

Alexander, Heather D., Courtney Siegert, J. Stephen Brewer, Jesse Kreye, Marcus A. Lashley, Jennifer K. McDaniel, Alison K. Paulson, Heidi J. Renninger, and J. Morgan Varner. "Mesophication of Oak Landscapes: Evidence, Knowledge Gaps, and Future Research." *BioScience* 71, no. 5 (May 1, 2021): 531–542. https://doi.org/10.1093/biosci/biaa169.

Anchor Point Media. "PTSD and Why I Had to Leave," with Mike West. *Anchor Point Podcast*, episode 72, May 3, 2021. https://anchorpointpodcast.com/episodes-1.

———. "Wildfire Fire Saved My Life," with Mark Booker. *Anchor Point Podcast*, episode 82, September 5, 2021. https://anchorpointpodcast.com/episodes-1.

Anderson, M. Kat. "Introduction to Omer C. Stewart's Article." *Fire Ecology* 10, no. 2 (August 2014): 1–3. https://doi.org/10.4996/fireecology.1002001.

Anderson, M. Kat. *Tending the Wild: Native American Knowledge and the Management of California's Natural Resources*. Berkeley: University of California Press, 2005.

Andrews, Jonathan. "From Stack-Firing to Pyromania: Medico-Legal Concepts of Insane Arson in British, US and European Contexts, c. 1800–1913. Part 1." *History of Psychiatry* 212, no. 3 (2010). https://journals.sagepub.com/doi/abs/10.1177/0957154x09349705.

"Arson Defendants and Their Counsel: Twelve Samarcand Girls Get State Prison Terms." *News and Observer*, May 21, 1931.

Associated Press. "Yosemite Ecologist Who Pioneered Prescribed Burns Dies. Why Washburn Fire Timing Was 'Poetry.'" August 13, 2022. https://apnews.com/article/science-fires-california-obituaries-climate-and-environment-b8cfa71cbaa39460b977ae72864f7f2f.

Azuzz, Elizabeth. "Rekindling Old Relationship: Learning from Traditional Cultural Fire Practices." Presented at the Wildfire in the West Workshop (virtual), April 23, 2021. www.ijnr.org/2021-wildfire-itinerary.

Baijnath-Rodino, Janine A., Shu Li, Alexandre Martinez, Mukesh Kumar, Lenya N. Quinn-Davidson, Robert A. York, and Tirtha Banerjee. "Historical Seasonal Changes in Prescribed Burn Windows in California." *Science of the Total Environment* 836 (August 2022): 155723. https://doi.org/10.1016/j.scitotenv.2022.155723.

Bailey, Jeremy. "Stop Focusing on Ignitions and Start Investing in a Prescribed Fire Workforce." *Fire Adapted Communities Learning Network* (blog), March 7, 2019. https://fireadaptednetwork.org/stop-focusing-on-ignitions-and-start-investing-in-a-prescribed-fire-workforce/.

Balice, Randy G., Kathryn D. Bennett, and Marjorie A. Wright. *Burn Severities, Fire Intensities, and Impacts to Major Vegetation Types from the Cerro Grande Fire*. Los Alamos National Laboratory, December 2004.

Barnes, Jeffrey K. *Natural History of the Albany Pine Bush: Albany and Schenectady Counties, New York, Including a Field Guide and Trail Map*. Albany: New York State Museum, 2003.

Beadel, H. L. "Fire Impressions." Presented at the 1st Tall Timbers Fire Ecology Conference, Tallahassee, FL, 1962.

Beck, Abaki. "The Yurok Tribe Is Using California's Carbon Offset Program to Buy Back Its Land." Greenbiz, May 5, 2021. www.greenbiz.com/article/yurok-tribe-using-californias-carbon-offset-program-buy-back-its-land.

Beirich, Heidi, and Mark Potok. "Countering Anti-Immigration Extremism: The Southern Poverty Law Center's Strategies." *City University of New York Law Review* 12, no. 2 (Summer 2009).

Belcher, Claire M., Margaret E. Collinson, and Andrew C. Scott. "A 450-Million-Year History of Fire." In *Fire Phenomena and the Earth System*, edited by Claire M. Belcher, 229–249. Hoboken, NJ: John Wiley and Sons, 2013.

Berger, Carrie, Lauren Grand, Stephen Fitzgerald, and Daniel Leavell. "Fire FAQs—What Is Fire Severity?" Oregon State University. Last reviewed October 2022. https://catalog.extension.oregonstate.edu/em9222/html.

Biggs, Michael. "Dying Without Killing: Self-Immolations, 1963–2002." In *Making Sense of Suicide Missions*, edited by Diego Gambetta. Oxford, UK: University of Oxford Press, 2005.

Blevins, David, and Michael P. Schafale. *Wild North Carolina: Discovering the Wonders of Our State's Natural Communities*. Chapel Hill: University of North Carolina Press, 2011.

Bond, W., and J. Keeley. "Fire as a Global 'Herbivore': The Ecology and Evolution of Flammable Ecosystems." *Trends in Ecology and Evolution* 20, no. 7 (July 2005): 387–394. https://doi.org/10.1016/j.tree.2005.04.025.

Bond, William J., and Andrew C. Scott. "Fire and the Spread of Flowering Plants in the Cretaceous." *New Phytologist* 188, no. 4 (2010): 1137–1150. https://doi.org/10.1111/j.1469-8137.2010.03418.x.

Bowman, David M. J. S., Grant J. Williamson, John T. Abatzoglou, Crystal A. Kolden, Mark A. Cochrane, and Alistair M. S. Smith. "Human Exposure and Sensitivity to Globally Extreme Wildfire Events." *Nature Ecology and Evolution* 1, no. 3 (February 6, 2017): 1–6. https://doi.org/10.1038/s41559-016-0058.

Boyd, Brian. *Vladimir Nabokov: The American Years*. Princeton, NJ: Princeton University Press, 1991.

Brenner, Jim, and Dale Wade. "Florida's Revised Prescribed Fire Law: Protection for Responsible Burners." Proceedings of Fire Conference 2000: First National Congress on Fire, Ecology, Prevention, and Management, Tall Timbers Research Station, Tallahassee, FL, 2003.

Buckland, Mollie Kathleen. "What Is a Megafire? Defining the Social and Physical Dimensions of Extreme U.S. Wildfires (1988–2014)." Master's thesis, University of Colorado Boulder, 2019.

Buckley, Thomas. *Standing Ground: Yurok Indian Spirituality, 1850–1990*. Berkeley: University of California Press, 2002.

Bureau of Indian Affairs, Office of Trust Services, Division of Forestry and Wildland Fire Management. "Fiscal Year 2015 Reserved Treaty Rights Lands Plan." 2015. www.bia.gov/sites/default/files/dup/assets/public/pdf/idc1-030969.pdf.

California Fire Science Consortium. "Ogle Defends 1920's 'Light Burning.'" Research Brief for Resource Managers, June 2013. https://static1.squarespace.com/static/545a90ede4b026480c02c5c7/t/55288edce4b05be8751e41a2/1428721372247/Lightburning7_Ogle-1920_CFSC_June-2013.pdf.

Calvanico, Jessica R. "Arson Girls, Match-Strikers, and Firestarters: A Reflection on Rage, Racialization, and the Carcerality of Girlhood." *Signs: Journal of Women in Culture and Society* 47, no. 2 (January 2022): 399–424. https://doi.org/10.1086/716654.

Carmack, George. "A Visit to a Tragic Scene: Where a Forest Died." *Albuquerque Tribune*, June 17, 1971.

Carmichael, Stephen W. "Finding the First Fires with Microscopes." *Microscopy Today* 12, no. 6 (November 2004): 3–7. https://doi.org/10.1017/S1551929500065895.

Catlin, George. *Prairie Meadows Burning*. 1832. Oil on canvas. Quotation is from the website's "Artwork Description." Smithsonian American Art Museum. Accessed March 26, 2023. https://americanart.si.edu/artwork/prairie-meadows-burning-4326.

Chazan, Michael. "Toward a Long Prehistory of Fire." *Current Anthropology* 58, no. S16 (August 2017): S351–359. https://doi.org/10.1086/691988.

Collins, Beryl Robichaud, and Karl H. Anderson. *Plant Communities of New Jersey: A Study in Landscape Diversity*. New Brunswick, NJ: Rutgers University Press, 1994.

Conservation International. "Center for Communities and Conservation." Accessed May 4, 2023. www.conservation.org/about/center-for-communities-and-conservation.

Correal, Annie. "'Very Real, and Very Haunting': Reporting on a Man Who Set Himself on Fire." New York Times, May 30, 2018. www.nytimes.com/2018/05/29/insider/david-buckel-self-immolation-prospect-park.html.

Covington, W.W., M.M. Moore, "Postsettlement Changes in Natural Fire Regimes and Forest Structure," *Journal of Sustainable Forestry*, Volume 2, Issue 1-2, 1994. https://doi.org/10.1300/J091v02n01_07.

Crawford, Matthew B. *The World Beyond Your Head: How to Flourish in an Age of Distraction*. New York: Farrar, Straus and Giroux, 2015.

Deloria, Vine, Jr. *Red Earth, White Lies: Native Americans and the Myth of Scientific Fact*. Golden, CO: Fulcrum Publishing, 1997.

Desmond, Matthew. *On the Fireline: Living and Dying with Wildland Firefighters*. Reprint ed. Chicago: University of Chicago Press, 2009.

Dey, Daniel, and Callie Schweitzer. "A Review on the Dynamics of Prescribed Fire, Tree Mortality, and Injury in Managing Oak Natural Communities to Minimize Economic Loss in North America." *Forests* 9, no. 8 (July 30, 2018): 461. https://doi.org/10.3390/f9080461.

Dillinger, Johannes. "Organized Arson as a Political Crime: The Construction of a 'Terrorist' Menace in the Early Modern Period." *Crime, History and Societies* 10, no. 2 (December 1, 2006): 101–121. https://doi.org/10.4000/chs.221.

Dyer, John. "Recoding the Chestnut." Boston.com, February 3, 2011. http://archive.boston.com/news/local/articles/2011/02/03/seven_communities_west_of_boston_host_orchards_devoted_to_reviving_american_chestnut_tree/.

Fei, Songlin, Ningning Kong, Kim C. Steiner, W. Keith Moser, and Erik B. Steiner. "Change in Oak Abundance in the Eastern United States from 1980 to 2008." *Forest Ecology and Management* 262, no. 8 (October 2011): 1370–1377. https://doi.org/10.1016/j.foreco.2011.06.030.

Fickle, James E. *Green Gold: Alabama's Forests and Forest Industries*. Tuscaloosa: University of Alabama Press, 2014.

Fimrite, Peter. "Yurok Indians Exult at Return of Sacred Cache." SFGate, August 13, 2010.

Flematti, Gavin R., Kingsley W. Dixon, and Steven M. Smith. "What Are Karrikins and How Were They 'Discovered' by Plants?" *BMC Biology* 13 (December 21, 2015): 108. https://doi.org/10.1186/s12915-015-0219-0.

Frechette, Alain, Chloe Ginsburg, and Wayne Walker. *A Global Baseline of Carbon Storage in Collective Lands: Indigenous and Local Community Contributions to Climate Change Mitigation*. Rights and Resources Initiative, September 2018. https://rightsandresources.org/wp-content/uploads/2018/09/A-Global-Baseline_RRI_Sept-2018.pdf.

Frierson, Cathy A. *All Russia Is Burning! A Cultural History of Fire and Arson in Late Imperial Russia*. Seattle: University of Washington Press, 2002.

Fromm, Michael, Daniel T. Lindsey, Rene Servranckx, Glenn K. Yum, Thomas Trickl, Robert J. Sica, Paul Doucet, and Sophie Godin-Beekman. "The Untold Story of Pyrocumulonimbus." *Bulletin of the American Meteorological Society* 91, no. 9 (2010): 1193–1209.

Gabbert, Bill. "Experienced Wildland Firefighter Explains Why He Resigned." *Wildfire Today*, April 15, 2022. https://wildfiretoday.com/2022/04/14/experienced-wildland-firefighter-explains-why-he-resigned/.

Gammage, Bill. *The Biggest Estate on Earth: How Aborigines Made Australia*. Reprint ed. Crows Nest, NSW: Allen and Unwin, 2013.

———. "Fire in 1788: The Closest Ally." *Australian Historical Studies* 42, no. 2 (2011). https://doi.org/10.1080/1031461X.2011.566273.

Geographic Area Coordination Centers. "National Year-to-Date Report on Fires and Acres Burned." Updated May 20, 2023. https://gacc.nifc.gov/sacc/predictive/intelligence/NationalYTDbyStateandAgency.pdf.

Glasspool, I. J., D. Edwards, and L. Axe. "Charcoal in the Silurian as Evidence for the Earliest Wildfire." *Geology* 32, no. 5 (May 1, 2004): 381–383. https://doi.org/10.1130/G20363.1.

Goldring, Winifred. "The Oldest Known Petrified Forest." *Scientific Monthly* 24, no. 6 (June 1927): 514–529.

Gowlett, J. A. J. "The Discovery of Fire by Humans: A Long and Convoluted Process." *Philosophical Transactions of the Royal Society B: Biological Sciences* 371, no. 1696 (June 5, 2016): 20150164. https://doi.org/10.1098/rstb.2015.0164.

Grassroots Wildland Firefighters. "Mike West Letter." Accessed March 7, 2023. www.grassrootswildlandfirefighters.com/mike-west-letter.

Greenberg, Cathryn H., Beverly S. Collins, Scott Goodrick, Michael C. Stambaugh, and Gary R. Wein. "Introduction to Fire Ecology Across USA Forested Ecosystems: Past, Present, and Future." In *Fire Ecology and Management: Past, Present, and Future of US Forested Ecosystems*, edited by Cathryn H. Greenberg and Beverly Collins, 39:1–30. Managing Forest Ecosystems. Cham: Springer International Publishing, 2021.

Hagmann, R. K., P. F. Hessburg, S. J. Prichard, N. A. Povak, P. M. Brown, P. Z. Fulé, R. E. Keane, E. E. Knapp, J. M. Lydersen, and K. L. Metlen. "Evidence for Widespread Changes in the Structure, Composition, and Fire Regimes of Western North American Forests." *Ecological Applications* 31, no. 8 (2021): e02431.

Halsey, Richard W. Letter to the editor, *Los Angeles Times*, August 2, 2022. www.latimes.com/opinion/letters-to-the-editor/story/2022-08-02/getting-indigenous-controlled-burning-all-wrong.

Hanh, Thich Nhat, Kosen Gregory Snyder, and Thomas Merton. *Vietnam: Lotus in a Sea of Fire: A Buddhist Proposal for Peace.* 2nd ed. Berkeley, CA: Parallax Press, 2022.

Hankins, Don, Scott Stephens, and Sara A. Clark. "Op-Ed: Why Forest Managers Need to Team Up with Indigenous Fire Practitioners." *Los Angeles Times*, July 31, 2022. www.latimes.com/opinion/story/2022-07-31/west-wildfire-prescribed-burn-indigenous-fire-practices.

Haraway, Donna J. *Staying with the Trouble: Making Kin in the Chthulucene.* Experimental Futures. Durham, NC: Duke University Press, 2016.

Harley, F. W. Letter to Forest Supervisor, "Klamath National Forest—Klamath Fires." January 30, 1918. www.karuk.us/images/docs/dnr/FWHarley_USFSRanger_1918_OrleansRD.pdf.

Harper, Roland M. *Economic Botany of Alabama.* University, AL: Geographical Survey of Alabama, 1913.

———. "Historical Notes on the Relation of Fires to Forests." Presented at the 1st Tall Timbers Fire Ecology Conference, Tallahassee, FL, 1962.

Hart, Justin L., Megan L. Buchanan, "History of Fire in Eastern Oak Forests and Implications for Restoration." Proceedings of the 4th Fire in Eastern Oak Forests Conference, Springfield, MO, May 17–19, 2011.

Haugo, Ryan D., Bryce S. Kellogg, C. Alina Cansler, Crystal A. Kolden, Kerry B. Kemp, James C. Robertson, Kerry L. Metlen, Nicole M. Vaillant, and

Christina M. Restaino. "The Missing Fire: Quantifying Human Exclusion of Wildfire in Pacific Northwest Forests, USA." *Ecosphere* 10, no. 4 (2019): e02702. https://doi.org/10.1002/ecs2.2702.

He, Tianhua, Claire M. Belcher, Byron B. Lamont, and Sim L. Lim. "A 350-Million-Year Legacy of Fire Adaptation Among Conifers." *Journal of Ecology* 104, no. 2 (2016): 352–363. https://doi.org/10.1111/1365-2745.12513.

He, Tianhua, and Byron B. Lamont. "Baptism by Fire: The Pivotal Role of Ancient Conflagrations in Evolution of the Earth's Flora." *National Science Review* 5, no. 2 (March 2018): 237–254.

He, Tianhua, Juli G. Pausas, Claire M. Belcher, Dylan W. Schwilk, and Byron B. Lamont. "Fire-Adapted Traits of Pinus Arose in the Fiery Cretaceous." *New Phytologist* 194, no. 3 (2012): 751–759. https://doi.org/10.1111/j.1469-8137.2012.04079.x.

Hoffman, Kira M., Ken P. Lertzman, and Brian M. Starzomski. "Ecological Legacies of Anthropogenic Burning in a British Columbia Coastal Temperate Rain Forest." *Journal of Biogeography* 44, no. 12 (2017): 2903–2915. https://doi.org/10.1111/jbi.13096.

Howell, Carol L. *Cannibalism Is an Acquired Taste: And Other Notes from Conversations with Anthropologist Omer C. Stewart.* Illustrated ed. Niwot: University Press of Colorado, 1998.

Hughes, Jim. "It's Rare for a 'Prescribed' Fire to Get Away." *Denver Post*, May 12, 2000.

Humphries, Aaron. "Wildland Fire Leadership: Mental Health and Being a Hotshot." *Wildland Fire Leadership* (blog), March 31, 2020. http://wildlandfireleadership.blogspot.com/2020/03/mental-health-and-being-hotshot.html.

Huntsinger, Lynn, and Sarah McCaffrey. "A Forest for the Trees: Forest Management and the Yurok Environment, 1850 to 1994." *American Indian Culture and Research Journal* 19, no. 4 (January 1, 1995): 155–192. https://doi.org/10.17953/aicr.19.4.cv0758kh373323h1.

Ingalsbee, Timothy. "Whither the Paradigm Shift? Large Wildland Fires and the Wildfire Paradox Offer Opportunities for a New Paradigm of Ecological Fire Management." *International Journal of Wildland Fire* 26, no. 7 (2017): 557. https://doi.org/10.1071/WF17062.

Jefferson, Lara, Marcello Pennacchio, and Kayri Havens-Young. *The Ecology of Plant-Derived Smoke: Its Use in Seed Germination.* Oxford, UK: Oxford University Press, 2014.

Jenkins, Jeffrey, Jan van Wagtendonk, and Mark Fincher. "The Evolution of Management Science to Inform Carrying Capacity of Overnight Visitor Use in the Yosemite Wilderness." *International Journal of Wilderness* 27, no. 2 (August 2021). https://ijw.org/evolution-management-science-carrying-capacity-overnight-visitors-yosemite-wilderness/.

Jose, Shibu, and Eric J. Jokela. *The Longleaf Pine Ecosystem: Ecology, Silviculture, and Restoration*. Berlin: Springer Science and Business Media, 2006.

Journals of the Lewis & Clark Expedition. "Wednesday, June 25, 1806." Accessed March 27, 2023. https://lewisandclarkjournals.unl.edu/item/lc.jrn.1806-06-25.

Jules, Erik S., Melissa H. DeSiervo, Matthew J. Reilly, Drew S. Bost, and Ramona J. Butz. "The Effects of a Half Century of Warming and Fire Exclusion on Montane Forests of the Klamath Mountains, California, USA." *Ecological Monographs* 92, no. 4 (2022): e1543. https://doi.org/10.1002/ecm.1543.

Jung, Yoohyun. "Not All Large Fires Are Bad. Here's Why the Burn Severity—Not Just Size—Matters." *San Francisco Chronicle*, July 25, 2022. www.sfchronicle.com/projects/2022/california-wildfire-severity-map/.

Kane, Jeffrey M., J. Morgan Varner, Michael C. Stambaugh, and Michael R. Saunders. "Reconsidering the Fire Ecology of the Iconic American Chestnut." *Ecosphere* 11, no. 10 (2020): e03267. https://doi.org/10.1002/ecs2.3267.

Kauffmann, Michael, and Justin Garwood. *The Klamath Mountains: A Natural History*. Kneeland, CA: Backcountry Press, 2022.

Knight, Clarke A., Charles V. Cogbill, Matthew D. Potts, James A. Wanket, and John J. Battles. "Settlement-Era Forest Structure and Composition in the Klamath Mountains: Reconstructing a Historical Baseline." *Ecosphere* 11, no. 9 (2020): e03250. https://doi.org/10.1002/ecs2.3250.

Knott, Jonathan A., Johanna M. Desprez, Christopher M. Oswalt, and Songlin Fei. "Shifts in Forest Composition in the Eastern United States." *Forest Ecology and Management* 433 (February 2019): 176–183. https://doi.org/10.1016/j.foreco.2018.10.061.

Kolden, Crystal A. "We're Not Doing Enough Prescribed Fire in the Western United States to Mitigate Wildfire Risk." *Fire* 2, no. 2 (June 2019): 30. https://doi.org/10.3390/fire2020030.

Komarek, E. V. "Fire—and the Ecology of Man." Presented at the 7th Tall Timbers Fire Ecology Conference, Hoberg, CA, 1967.

———. "The Use of Fire: An Historical Background." Presented at the 1st Tall Timbers Fire Ecology Conference, Tallahassee, FL, 1962.

Kosek, Jake. *Understories: The Political Life of Forests in Northern New Mexico*. Durham, NC: Duke University Press, 2006.

Kroeber, A. L., and E. W. Gifford. *World Renewal: A Cult System of Native Northwest California*. Anthropological Records, vol. 13. Berkeley: University of California Press, 1949.

Kudish, Michael. *The Catskill Forest: A History*. Fleischmanns, NY: Purple Mountain Press, 2000.

Kupfer, John A., Adam J. Terando, Peng Gao, Casey Teske, J. Kevin Hiers. "Climate Change Projected to Reduce Prescribed Burning Opportunities in the South-Eastern United States." *International Journal of Wildland Fire* 29, no. 9 (June 1, 2020): 764–778. https://doi.org/10.1071/WF19198.

Latham, Den, and Shibu Jose. *Painting the Landscape with Fire: Longleaf Pines and Fire Ecology*. Columbia: University of South Carolina Press, 2013.

Leopold, A. S., S. A. Cain, C. M. Cottam, I. N. Gabrielson, and T. L. Kimball. *Wildlife Management in the National Parks*. US Department of the Interior, National Park System, March 4, 1963.

Levy, Sharon. "Rekindling Native Fires." *BioScience* 55, no. 4 (April 1, 2005): 303–308. https://doi.org/10.1641/0006-3568(2005)055[0303:RNF]2.0.CO;2.

Lewis, Henry T. "Ecological and Technological Knowledge of Fire: Aborigines Versus Park Rangers in Northern Australia." *American Anthropologist* 91, no. 4 (1989): 940–961.

Lewis, Henry T., and Theresa A. Ferguson. "Yards, Corridors, and Mosaics: How to Burn a Boreal Forest." *Human Ecology* 16, no. 1 (March 1, 1988): 57–77. https://doi.org/10.1007/BF01262026.

Lewis, Nolan D. C., and Helen Yarnell. *Pathological Firesetting; Pyromania*. New York: Nervous and Mental Disease Monographs, 1951.

Lunder, Zeke. "Dixie Fire—8/4/2021—Evening." *The Lookout* (blog), August 4, 2021. https://the-lookout.org/2021/08/04/dixie-fire-8-4-2021-evening/.

———. "Dixie Fire—8/5/2021—Afternoon." *The Lookout* (blog), August 5, 2021. https://the-lookout.org/2021/08/05/dixie-fire-8-5-2021-afternoon-update/.

———. "Dixie Fire—Firing Operations." *The Lookout* (blog), September 1, 2021. https://the-lookout.org/2021/09/01/dixie-fire-firing-operations/.

———. "Dixie Fire—Friday the 13th." *The Lookout* (blog), August 12, 2021. https://the-lookout.org/2021/08/12/8-12-2021-evening/.

———. "Dixie Fire Severity—8/12/2021." *The Lookout* (blog), August 13, 2021. https://the-lookout.org/2021/08/13/dixie-fire-severity-8-12-2021/.

MacDonald, Katharine, Fulco Scherjon, Eva van Veen, Krist Vaesen, and Wil Roebroeks. "Middle Pleistocene Fire Use: The First Signal of Widespread Cultural Diffusion in Human Evolution." *Proceedings of the National Academy of Sciences* 118, no. 31 (August 3, 2021): e2101108118. https://doi.org/10.1073/pnas.2101108118.

Maclean, Norman. *Young Men and Fire*. Chicago: University of Chicago Press, 1992.

Maguire, Dannielle. "Bushfire Destroys Couple's North Rosedale Home but Climate Change Sign Survives." ABC News, January 3, 2020. www.abc.net.au/news/2020-01-04/north-rosedale-bushfire-guts-home-climate-change-sign-survives/11839642.

Marder, Michael. *Green Mass: The Ecological Theology of St. Hildegard of Bingen*. Stanford, CA: Stanford University Press, 2021.

Marks-Block, Tony, and William Tripp. "Facilitating Prescribed Fire in Northern California Through Indigenous Governance and Interagency Partnerships." *Fire* 4, no. 3 (September 2021): 37. https://doi.org/10.3390/fire4030037.

McCovey, Mavis, and John F. Salter. *Medicine Trails: A Life in Many Worlds*. Berkeley, CA: Heyday Books, 2009.

McGranahan, Devan Allen, and Carissa L. Wonkka. *Ecology of Fire-Dependent Ecosystems: Wildland Fire Science, Policy, and Management*. Boca Raton, FL: CRC Press, 2020.

McRae, Rick. "An Error of Definition—and a Need to Make Valid Science Key to Our Best Practices." *Wildfire Magazine*, 2019. www.iawfonline.org/article/dialogue-issue-paper-extreme-fires/.

Meyer, Marc D., and Hugh D. Safford. "Giant Sequoia Regeneration in Groves Exposed to Wildfire and Retention Harvest." *Fire Ecology* 7, no. 2 (August 1, 2011): 2–16. https://doi.org/10.4996/fireecology.0702002.

Meyer-Berthaud, B., and A.-L. Decombeix. "Palaeobotany: In the Shade of the Oldest Forest." *Nature* (February 29, 2012). https://doi.org/10.1038/483041a.

Michelsen, Nicholas. "The Political Subject of Self-Immolation." In *Occupying Subjectivity: Being and Becoming Radical in the 21st Century*, edited by Chris Rossdale. New York: Routledge, 2017.

Mittelbach, Margaret, and Michael Crewdson. "Chasing Nabokov's Elusive and Endangered True Love." *New York Times*, July 14, 2000. www.nytimes.com/2000/07/14/arts/chasing-nabokov-s-elusive-and-endangered-true-love.html.

"Montana and California Go Big, Dirty August Started 2 Days Early." *Hotshot Wake Up* (blog), July 30, 2022. https://thehotshotwakeup.substack.com/p/montana-and-california-go-big-dirty.

Muir, John. *Our National Parks*. Berkeley: University of California Press, 1991.

———. *The Mountains of California*. San Francisco: Sierra Club Books, 1989.

Murie, James R., and Douglas R. Parks. *Ceremonies of the Pawnee, Part I: The Skiri*. Washington, DC: Smithsonian Contributions to Anthropology 27, 1981.

Nabokov, Vladimir. *Pnin*. Reissue ed. New York: Vintage, 1989.

National Wildfire Coordinating Group. "10 Standard Firefighting Orders, PMS 110." Last reviewed January 25, 2023. www.nwcg.gov/publications/pms110.

———. "2022 WOR Day 7: South Canyon Fire (Colorado)—July 6, 1994," October 2022. www.nwcg.gov/committee/6mfs/weekremembrance/wor-2022-day7.

———. "Fight Fire Aggressively, Having Provided for Safety First." Last modified March 2022. www.nwcg.gov/committee/6mfs/fight-fire-aggressively.

———. "Module 1: Basic Concepts of Wildland Fire." Accessed May 17, 2023. https://training.nwcg.gov/classes/S190/508Files/071231_s190_m1_508.pdf.

———. "S-190 Introduction to Wildland Fire Behavior, 2020." Last reviewed December 16, 2022. www.nwcg.gov/publications/training-courses/s-190.

———. "WFSTAR: Alabaugh Fire Deployment Story (MTDC)." Video, YouTube, 2018. www.youtube.com/watch?v=F4ILBdCUBR4.

Bibliography

———. "WFSTAR: Holloway Fire Deployment Deployment Story (MTDC)." Video, YouTube. 2018. www.youtube.com/watch?v=pnfieTf0EZY.

Nature Communications. "Climate-Induced Variations in Global Wildfire Danger from 1979 to 2013." July 14, 2015. www.nature.com/articles/ncomms8537.

Nature Conservancy. "Through the Fire: Restoring Forest Resilience." Video, YouTube. 2022. www.youtube.com/watch?v=uhVt1pPFMgI.

Nijhuis, Michelle. "The West Coast Wildfires Are Apocalypse, Again." *New Yorker*, October 19, 2020. www.newyorker.com/news/annals-of-a-warming-planet/the-west-coast-wildfires-are-apocalypse-again.

Norgaard, Kari Marie. *Salmon and Acorns Feed Our People: Colonialism, Nature, and Social Action*. Illustrated ed. New Brunswick, NJ: Rutgers University Press, 2019.

North, M. P., R. A. York, B. M. Collins, M. D. Hurteau, G. M. Jones, E. E. Knapp, L. Kobziar, et al. "Pyrosilviculture Needed for Landscape Resilience of Dry Western United States Forests." *Journal of Forestry*, no. fvab026 (May 21, 2021). https://doi.org/10.1093/jofore/fvab026.

NPS Western Regions Wildland Fire Resource Advisor Course (N-9042). Core Presentations, 2022.

Nussbaum, Martha C. "A Peopled Wilderness." *New York Review of Books*, December 8, 2022. www.nybooks.com/articles/2022/12/08/a-peopled-wilderness-martha-c-nussbaum/.

"Nuttall Fire Series: Thomas Thom, Flagstaff IHC Sawyer." *Wildland Fire Lessons Learned Center* podcast, July 4, 2017. https://wildfirelessons.podbean.com/e/nuttall-fire-series-thomas-Thom-flagstaff-ihc-sawyer/.

Parks, Sean A., Carol Miller, Marc-André Parisien, Lisa M. Holsinger, Solomon Z. Dobrowski, and John Abatzoglou. "Wildland Fire Deficit and Surplus in the Western United States, 1984–2012." *Ecosphere* 6, no. 12 (December 2015): art275. https://doi.org/10.1890/ES15-00294.1.

Patterson, William A., and Kenneth E. Sassaman. "Indian Fires in the Prehistory of New England." In *Holocene Human Ecology in Northeastern North America*, edited by George P. Nicholas, 107–135. Interdisciplinary Contributions to Archaeology. Boston: Springer US, 1988. https://doi.org/10.1007/978-1-4899-2376-9_6.

Pausas, Juli G., and Jon E. Keeley. "A Burning Story: The Role of Fire in the History of Life." *BioScience* 59, no. 7 (July 1, 2009): 593–601. https://doi.org/10.1525/bio.2009.59.7.10.

Pellatt, Marlow G., and Ze'ev Gedalof. "Environmental Change in Garry Oak (Quercus Garryana) Ecosystems: The Evolution of an Eco-Cultural Landscape." *Biodiversity and Conservation* 23, no. 8 (July 1, 2014): 2053–2067. https://doi.org/10.1007/s10531-014-0703-9.

Pellegrini, Adam F. A., Jennifer Harden, Katerina Georgiou, Kyle S. Hemes, Avni Malhotra, Connor J. Nolan, and Robert B. Jackson. "Fire Effects on the Persistence of Soil Organic Matter and Long-Term Carbon Storage." *Nature Geoscience* 15, no. 1 (January 2022): 5–13. https://doi.org/10.1038/s41561-021-00867-1.

Peterson, David A., Michael D. Fromm, Richard H. D. McRae, James R. Campbell, Edward J. Hyer, Ghassan Taha, Christopher P. Camacho, George P. Kablick, Chris C. Schmidt, and Matthew T. DeLand. "Australia's Black Summer Pyrocumulonimbus Super Outbreak Reveals Potential for Increasingly Extreme Stratospheric Smoke Events." *NPJ Climate and Atmospheric Science* 4, no. 1 (July 13, 2021): 1–16. https://doi.org/10.1038/s41612-021-00192-9.

Petty, John. "Help Came Quickly in Jemez Fire." *Albuquerque Journal*, June 12, 1971.

Phillips, Tony. "The Tunguska Impact: 100 Years Later." NASA Science: Share the Science, June 30, 2008. https://science.nasa.gov/science-news/science-at-nasa/2008/30jun_tunguska.

Porter, Frank W. *The Art of Native American Basketry: A Living Legacy*. Westport, CT: Greenwood Publishing, 1990.

Poulos, Helen. "Fire in the Northeast: Learning from the Past, Planning for the Future." *Journal of Sustainable Forestry* 34, no. 1–2 (February 17, 2015): 6–29. https://doi.org/10.1080/10549811.2014.973608.

Power, Mitchell J. "A 21,000-Year History of Fire." In *Fire Phenomena and the Earth System*, edited by Claire M. Belcher, 207–227. Hoboken, NJ: John Wiley and Sons, 2013. https://doi.org/10.1002/9781118529539.ch11.

Pyne, Stephen J. *Between Two Fires: A Fire History of Contemporary America*. Tucson: University of Arizona Press, 2015.

———. "Commentaries, Otherwise Unpublished." *Stephen J. Pyne* (blog), July 1, 2017. www.stephenpyne.com/blog/posts/27919.

———. *Fire in America: A Cultural History of Wildland and Rural Fire*. Paperback ed. Seattle: University of Washington Press, 2017.

———. *Fire on the Rim: A Firefighter's Season at the Grand Canyon*. Seattle: University of Washington Press, 2017.

———. *Florida: A Fire Survey*. Tucson: University of Arizona Press, 2016.

———. *The Pyrocene: How We Created an Age of Fire, and What Happens Next*. Oakland: University of California Press, 2021.

———. *Tending Fire: Coping with America's Wildland Fires*. Washington, DC: Island Press, 2004.

———. *World Fire: The Culture of Fire on Earth*. Seattle: University of Washington Press, 1997.

Quillen, Ed. "Change We Could Believe In." *High Country News*, December 22, 2008. www.hcn.org/issues/40.23/change-we-could-believe-in.

Bibliography

"Remembering Jan van Wagtendonk, Who Shaped Fire Management in Yosemite National Park." Wildfire Today, August 14, 2022. https://wildfiretoday.com/2022/08/14/remembering-jan-van-wagtendonk-who-shaped-fire-management-in-yosemite-national-park/.

Rights and Resources. "New Analysis Reveals That Indigenous Peoples and Local Communities Manage 300,000 Million Metric Tons of Carbon in Their Trees and Soil—33 Times Energy Emissions from 2017." September 9, 2018. https://rightsandresources.org/blog/new-analysis-reveals-that-indigenous-peoples-and-local-communities-manage-300000-million-metric-tons-of-carbon-in-their-trees-and-soil-33-times-energy-emissions-from-2017/.

Rooney, B. "Burnin' Bill and the Dixie Crusaders." *American Forests* 99, no. 5/6 (1993): 35.

Roosevelt, Theodore. Letter to John Muir. May 19, 1903. Online Archive of California, accessed March 14, 2023. https://oac.cdlib.org/ark:/13030/kt3199r8rw/?order=2.

Russell, Gregory, Joseph G. Champ, David Flores, Michael Martinez, Alan M. Hatch, Esther Morgan, and Paul Clarke. "Doing Work on the Land of Our Ancestors: Reserved Treaty Rights Lands Collaborations in the American Southwest." *Fire* 4, no. 1 (March 2021): 7. https://doi.org/10.3390/fire4010007.

Sabalow, Ryan, and Dale Kasler. "Wildfire Experts Escalate Fight over Saving California Forests." *Sacramento Bee*, October 25, 2021.

Salmon, Enrique. "Kincentric Ecology: Indigenous Perceptions of the Human-Nature Relationship." *Ecological Applications* 10 (October 1, 2000): 1327–1332. https://doi.org/10.2307/2641288.

Schroeder, Mark J., and Charles C. Buck. *Fire Weather, PMS 426-1: A Guide for Application of Meteorological Information to Forest Fire Control Operations.* US Department of Agriculture Forest Service, May 1970.

Scudder, Samuel Hubbard. *The Butterflies of the Eastern United States and Canada: With Special Reference to New England.* Cambridge, MA: The author, 1889.

Shea, John P. *Getting at the Roots of Man-Caused Forest Fires: A Case Study of a Southern Forest Area.* United States Department of Agriculture Forest Service, US Government Printing Office, 1940.

Sobrevila, Claudia. *The Role of Indigenous Peoples in Biodiversity Conservation: The Natural but Often Forgotten Partners.* World Bank, May 2008.

Stein, William E., Christopher M. Berry, Jennifer L. Morris, Linda VanAller Hernick, Frank Mannolini, Charles Ver Straeten, Ed Landing, et al. "Mid-Devonian Archaeopteris Roots Signal Revolutionary Change in Earliest Fossil Forests." *Current Biology* 30, no. 3 (February 3, 2020): 421–431.e2. https://doi.org/10.1016/j.cub.2019.11.067.

Stephens, Scott. "Fire on the Mountain (and Elsewhere): Forests, Ecology, and What Does 'Management' Really Mean?" Presented at the Wildfire in the West Workshop (virtual), April 23, 2021.

Stephenson, Nathan L. "Reference Conditions for Giant Sequoia Forest Restoration: Structure, Process, and Precision," *Ecological Applications*, Volume 9, Issue 30 (November 1999): https://doi.org/10.1890/1051-0761(1999)009 [1253:RCFGSF]2.0.CO;2.

Stephenson, Nathan L., and Christy Brigham. "Preliminary Estimates of Sequoia Mortality in the 2020 Castle Fire," National Park Service, June 25, 2021. https://www.nps.gov/articles/000/preliminary-estimates-of-sequoia-mortality-in-the-2020-castle-fire.htm

Stewart, Omer C. "Barriers to Understanding the Influence of Use of Fire by Aborigines on Vegetation." Presented at the 2nd Tall Timbers Fire Ecology Conference, Tallahassee, FL, 1962.

———. *Forgotten Fires: Native Americans and the Transient Wilderness*. Edited by Henry T. Lewis and M. Kat Anderson. Illustrated ed. Norman: University of Oklahoma Press, 2009.

Stoddard, H. L., Sr. "Use of Fire in Pine Forests and Game Lands of the Deep Southeast." Presented at the 1st Tall Timbers Fire Ecology Conference, Tallahassee, FL, 1962.

Stowe, Johnny P. "The Role of Ritual and Ceremony in Wildlands Conservation: Reestablishing Primal Connections." Proceedings of the Sixth Longleaf Alliance Regional Conferences, Tipton, GA, November 13–16, 2006.

Struzik, Edward. *Firestorm: How Wildfire Will Shape Our Future*. Illustrated ed. Washington, DC: Island Press, 2017.

Sullivan, Alan P., and Kathleen M. Forste. "Fire-Reliant Subsistence Economies and Anthropogenic Coniferous Ecosystems in the Pre-Columbian Northern American Southwest." *Vegetation History and Archaeobotany* 23, no. 1 (May 1, 2014): 135–151. https://doi.org/10.1007/s00334-014-0434-6.

Swetnam, Thomas W. "Fire History and Climate Change in Giant Sequoia Groves." *Science* 262, no. 5135 (November 5, 1993): 885–889. https://doi.org/10.1126/science.262.5135.885.

Taylor, Thomas. "Hello Darkness...." *Wildland Fire Lessons Learned Center* (blog), February 16, 2021. https://wildfirelessons.blog/2021/02/16/hello-darkness/.

———. "Is LCES Dead?" *Wildland Fire Lessons Learned Center* (blog), June 16, 2020, https://wildfirelessons.blog/2020/06/16/is-lces-dead/.

Thompson, Glen. "Fire and the Environment—Application Section." Presented at the 1st Tall Timbers Fire Ecology Conference, Tallahassee, FL, 1962.

Thompson, Lucy. *To the American Indian*. Berkeley, CA: Heyday Books, 1991.

Thoreau, Henry David. "Thoreau's Journal (Part II)." *The Atlantic*, February 1905.

Bibliography

Thoreau, Henry David, and Edward Hoagland. *The Maine Woods.* New York: Penguin Classics, 1988.

Torero, Jose L. "An Introduction to Combustion in Organic Materials." In *Fire Phenomena and the Earth System,* edited by Claire M. Belcher, 1–13. Hoboken, NJ: John Wiley and Sons, 2013.

Trapani, Ryan. "First Growth Forest in the Catskills with Forest Historian Dr. Michael Kudish." *From the Forest* podcast, September 22, 2022. https://fromtheforest.podbean.com.

Treuer, David. "Return the National Parks to the Tribes." *The Atlantic,* April 12, 2021. www.theatlantic.com/magazine/archive/2021/05/return-the-national-parks-to-the-tribes/618395/.

Tripp, Bill. "Cultural Resource Message." Klamath TREX, October 13, 2022.

Tsing, Anna Lowenhaupt, Heather Anne Swanson, Elaine Gan, and Nils Bubandt. *Arts of Living on a Damaged Planet: Ghosts of the Anthropocene.* Minneapolis: University of Minnesota Press, 2017.

United States Forest Service. *Nuttall Complex Fire Shelter Deployment Review: Factual Report and Management Evaluation Report.* December 6, 2004.

———. "Statement from Chief Randy Moore on Hermit's Peak Fire Review." June 20, 2022. www.fs.usda.gov/news/releases/statement-chief-randy-moore-hermit-fire.

———. *Thirtymile Fire Investigation: Accident Investigation Factual Report and Management Evaluation Report: Chewuch River Canyon, Winthrop, Washington.* September 2001.

van Wagtendonk, Jan W. "The History and Evolution of Wildland Fire Use." *Fire Ecology* 3, no. 2 (December 2007): 3–17. https://doi.org/10.4996/fireecology.0302003.

Viktora, Alex. "'Student of Fire': What Does It Really Mean?" *Two More Chains,* Winter 2017.

Washburn, Dorothy K. "Symmetry Analysis of Yurok, Karok, and Hupa Indian Basket Designs." *Empirical Studies of the Arts* 4, no. 1 (January 1, 1986): 19–45. https://doi.org/10.2190/VKF2-HVHH-X8RB-QT85.

Way, Albert G., Leon Neel, and Paul S. Sutter. *The Art of Managing Longleaf: A Personal History of the Stoddard-Neel Approach.* Athens: University of Georgia Press, 2012.

Weir, John R. *Conducting Prescribed Fires: A Comprehensive Manual.* Illustrated ed. College Station: Texas A&M University Press, 2009.

Weiser, Andrea, and Dana Lepofsky. "Ancient Land Use and Management of Ebey's Prairie, Whidbey Island, Washington." *Journal of Ethnobiology* 29, no. 2 (September 2009): 184–212. https://doi.org/10.2993/0278-0771-29.2.184.

Werth, Paul, Brian Potter, Craig Clements, Mark Finney, Jayson Forthofer, Sara McAllister, Scott Goodrick, Martin Alexander, and Miguel Cruz. *Synthesis of Knowledge of Extreme Fire Behavior: Volume I for Fire Managers.* US Forest Service, 2011. https://digitalcommons.unl.edu/jfspsynthesis/6.

Wheeler, Ben, ed. *Prescribed Fire in Nebraska: A Guide for Planning, Preparing and Conducting Your Prescribed Fire.* Pheasants Forever, 2017.

Whyte, Kyle Powys. "Our Ancestors' Dystopia Now: Indigenous Conservation and the Anthropocene." In *The Routledge Companion to the Environmental Humanities,* edited by Ursula Heise, Jon Christensen, and Michelle Niemann, 205–215. London: Routledge, 2017.

Wildland Fire Jobs. "How to Have a Successful Season." December 3, 2018. https://wildlandfirejobs.com/how-to-have-a-successful-season/.

WildlandFireLLC. "30 Mile Fire Fatality Case Study." Video, YouTube. 2012. www.youtube.com/watch?v=KK4SjPqYor4.

———. "Inside the Shelter." Video, YouTube. 2014. www.youtube.com/watch?v=mpS-Dpu8jvs.

———. "Nuttall Fire Episode One." Video, YouTube. 2017. www.youtube.com/watch?v=ofksOghWlg0.

Williams, Gerald W. "Wildland Fire Management in the 20th Century." *Fire Management Today,* Winter 2000.

Williams, Terry Tempest. "The Liturgy of Home." *Harvard Divinity Bulletin,* Spring/Summer 2018.

Williamson, Bhiamie, Jessica Weir, and Vanessa Cavanagh. "Strength from Perpetual Grief: How Aboriginal People Experience the Bushfire Crisis." The Conversation, January 9, 2020. https://theconversation.com/strength-from-perpetual-grief-how-aboriginal-people-experience-the-bushfire-crisis-129448.

Yeater, Ralph F. "Incendiarism: Its Cause and Prevention." Bachelor of Science thesis, Oregon State College, 1940.

"Yosemite Logging Unprecedented for a National Park, Says Conservation Group Suing to Stop It." *Fresno Bee,* June 15, 2022. www.fresnobee.com/news/california/yosemite/article262477057.html.

Yurkonis, Kathryn A., Josie Dillon, Devan A. McGranahan, David Toledo, and Brett J. Goodwin. "Seasonality of Prescribed Fire Weather Windows and Predicted Fire Behavior in the Northern Great Plains, USA." *Fire Ecology* 15, no. 1 (March 25, 2019): 7. https://doi.org/10.1186/s42408-019-0027-y.

Zimmer, Dieter. "Butterflies and Moths Named by and for Nabokov." A Guide to Nabokov's Butterflies and Moths, web version, 2012. www.d-e-zimmer.de/eGuide/PageOne.htm.

Zipf, Karin Lorene. *Bad Girls at Samarcand: Sexuality and Sterilization in a Southern Juvenile Reformatory.* Baton Rouge: Louisiana State University Press, 2016.

INDEX

Abernathy, Margaret, 81–82
Aboriginal Noongar language, 74–75
Aboriginals/fire, 2, 4
Adams, Ansel, 190
Africa and early ancestors/fire, 49
Agee, Jim, 262–263
Ahwahnechee Native American communities, 89–90, 189
Air Quality Index, 148
Alabama-Coushatta Tribe, Texas, 285
Alaska and early ancestors/fire, 49
Albany Pine Bush/fire
  description/history, 66, 68–70, 72
  ecological benefits, 85–86
  fire crew work, 71–74, 75
  O'Connor, 66, 71–74, 75
  plants/nutrients, 68–69
  See also pine barrens, New Jersey/New York
All Hands All Lands, Forest Stewards Guild, 233, 234
Amah Mutsun Tribal Band, California, 10
American chestnut tree (*Castanea dentata*), 50–51, 250
American Eugenics Society, 90
American Forestry Congress, 253

anchor point described, 21
Anchor Point, The podcast, 112
Anderson, M. Kat, 56–57, 60–61, 216, 290
angiosperms/grass angiosperms explosions, 5
anti-immigration policy, 95
Apache, 205, 240
Apache-Sitgreaves National Forest, Arizona, 223
Apalachicola National Forest, Florida, 4–5
Arab Spring, 84
Arbor Day, 25
Archaeopteris, 47
Arrowhead Hotshots, 193, 197
arsonists. *See* pyromaniacs
Art of Managing Longleaf, The (Neel), 261
"asbestos forest," 51
Ash Wednesday fires, Australia (1983), 4
asteroids
  fire ignition, 44–45
  Tunguska asteroid/consequences, 44–45
Atlantic, The, 305
August Complex/gigafire, California (2020), 37, 106, 167

Australia
  Aboriginals/fire, 2, 4
  Black Summer (2019-2020), 2–3, 4, 100
  fire, 1–4, 6, 100, 259
  forest descriptions, 1–2, 4
Azuzz, Elizabeth
  background, 61, 281
  fire, 61–62, 285–287, 292, 294, 302–303

backfiring described, 19
Bad Girls at Samarcand: Sexuality and Sterilization in a Southern Juvenile Reformatory (Zipf), 81–82
Bailey, Jeremy
  background/fire experience, 63–64, 202
  description/traits, 63
Bambi, 94
Bandelier National Monument, New Mexico, 241
Barnes, Jeffrey, 68, 69
Bauer, Justin, 225, 312, 313, 314
Beadel, Henry
  brother and Charlie, 252
  fire, 252–253, 261
Beckwourth Complex, 144
Between Two Fires: A Fire History of Contemporary America (Pyne), 198, 200
Biden, President, mandate on conservation, 315
Big Blowup (1910), 142
Bigfoot country, 169, 288
Biggest Estate on Earth, The (Gammage), 2
Biggs, Michael, 84
Big Sur fire, California (2013), 134
Bingen, Hildegard von, 44
BioScience, 290
Biswell, Harold, 106–107, 199–200, 262, 263
"black" described, 20
Blackfoot Watershed, Montana, 64–65
Black Hills, South Dakota, 311
Black Saturday bushfires, Australia (2009), 6

Black Summer, Australia (2019–2020), 2–3, 4, 100
Blodgett Experimental Forest, west of Lake Tahoe, 105
Blue Jay Fire, Yosemite National Park (2020), 195
Blue Ridge Mountains, 92–93
Bobwhite Quail, The: Its Habits, Preservation and Increase (Stoddard), 251–252
Boise National Forest. *See* Pioneer Fire, Boise National Forest
Bolomor Cave, Spain, 48
Bootleg Fire, Oregon, 140–141
boreal forest, Canada, 60
Bouazizi, Mohamed, 84
Brigham, Christy, 210, 211–212
British/colonies
  fire suppression, 87–88
  locals and, 88
Buckel, David, 84
Buckley, Dan
  background/fire experience, 194–198, 200, 201, 208, 209, 318, 319
  prescribed burns/letting fires burn, 195–196, 200, 201, 208
  Yosemite/Pyne, 194–198, 203
Buckley, Thomas, 275, 279, 283
Buddhists and self-immolation, 84–85
Bureau of Indian Affairs, 305–306
Bureau of Land Management, US, 18
Burma and fire, 88
Bursera trees, 75
butenolides, 74–75
Butterflies of the Eastern United States and Canada, The (Scudder), 67
Butterfly Station, Karner, New York, 67

Camp Fire/deaths, California (2018), 39, 136, 142, 146, 165–166
Carlson, Thea Maria
  background/fire experience, 296–298
  description, 295
  intentional community of, 296
Carr Fire, California, 102

Index

Carrick, Patty
  background/fire experience, 15, 16–17, 18, 29, 30, 35–36, 315, 320, 321, 323–324
  description/traits, 15–16, 302–303, 323–324
  Joe (husband), 16
Carson National Forest. *See* Ensenada RX prescribed burn (Carson National Forest)
Castanea dentata (American chestnut tree), 50–51, 250
Castle Fire, California (2020), 207, 212, 216
Catlin, George, 55
Catskill State Park, New York
  O'Connor, 45, 51–52
  tree species, 50–51
  woods/description, 45, 49–52
Cebollita Fire, New Mexico, 97–98
cedars, eastern redcedar (*Juniperus virginiana*), 25–26
cedars, western red cedar, 59
Ceremonies of the Pawnee (Murie), 34
Cerro Grande Fire, New Mexico (2000)
  original goal/ecological benefits, 242
  as prescribed burn/escape, 200, 241–242
  as social disaster, 200, 242–243
chainsaw course
  Eric/Justin, 312, 313, 314
  O'Connor, 311–314
Chapman, Neil
  background/fire experience, 220, 224–225
  prescribed fire, 220
  restoration project, 220, 224–225
charcoal
  description, 43
  preservation/knowledge from, 43, 49
Chazan, Michael, 48–49
chestnut tree (American chestnut tree, *Castanea dentata*), 50–51, 250
Cibola National Forest, 234
Citizen Potawatomi Nation, 307

Civilian Conservation Corps, 251
Clarke-McNary Act (1931), 252
classes on fires
  diversity/accessibility, 201–202
  O'Connor, 14, 23–24
  See also chainsaw course
climate change
  Buckel/self-immolation, 84
  carbon dioxide increase/statistics, 204, 215
  deniers-environmental activists against forest management comparison, 218
  Dixie Fire, California (2021) and, 144–145, 148, 168, 171
  giant sequoias/groves, 215–216
  Intergovernmental Panel on Climate Change reports/causes, 215
  Klamath region, 282–283
  response options, 216
  views of, 306–307
climate change/fires
  droughts, 3, 97, 100, 104, 105, 144–145, 168, 190, 200, 201, 208, 209, 214, 215, 216, 220, 242, 243, 251, 257, 315, 316–317
  fire effects on, 185
  fire paradox, 8–9
  megafires, 8, 101, 104–105, 107
  prescribed fires, 107, 193, 298, 316–317
  Pyne on, 185–186
  pyroCbs, 112
  restoration project/Flagstaff area, 221
  Stephenson on, 215–216
  wildfires as climate apocalypse symbol, 1
climax forest, 49–50
Clovis people, 246
Coconino National Forest, Arizona, 221, 223
cold-trailing (mop-up) described, 152
"Cold War on fire," 95
Cooperative Forest Fire Prevention Program, 94
Coronado National Forest, Arizona, 116

"corridors"/"fire corridors," 60
Crawford, Matthew B., 188
Cresta Dam, California, 135–136
Cretaceous period, 5, 75
Crime, History and Societies (journal), 83
Croatan National Forest, North Carolina, 263
cultural anthropology, 52
cultural landscapes, 53
cultural tree, 239–240
cup trenching, 153, 169

data and fires, 103
Deloria, Vine, Jr., 52
dendrochronologists described, 58
Denisovans, 47
Denver Post, 243
Department of Agriculture, US, 16
Department of Natural Resources, South Carolina, 248
Desmond, Matthew, 78
Devonian period, 45, 47
DeVries, David, 65
Diagnostic and Statistical Manual of Mental Disorders, 128–129
Diné people (Navajo), 225, 226, 228, 240, 321
"dirty August," 145
Disney, 94
Dixie Crusaders, 92
Dixie Fire, California (2021)
    arsonist, 147
    costs to fight, 186, 243
    crew sleep, 146, 154–155, 170, 182, 183
    descriptions/fire work, 135–137, 139–140, 141–143, 144–145, 147–155, 158–159, 160–163, 167–179, 182–183, 186–187, 193
    dirt biker and, 167–168
    dryness/climate change, 144–145, 148, 168, 171
    Frank/background, 146–147, 150, 153, 157, 177, 179–182
    "ghost" wildland firefighter and, 175–176
    Greenville, 139–140, 160–163, 167
    home protection work, 147–149
    hotshot crews, 143, 151, 174–175, 183, 186
    mop-up/cold-trailing, 152–154, 159, 169, 170, 171–172
    O'Connor, 137, 139, 141, 142–155, 157, 158–160, 169–172, 173–174, 175, 176–179, 182–183, 188
    origins, 135–136, 146
    prison camp training/crew members, 147, 150–151, 160, 180, 317
    pulling hose, 172
    pyroCbs, 136, 141
    "Pyrocene," 186
    See also specific individuals
Dodge, Wag, 21–22
downdrafts and pyroCbs, 111–112
"dragon eggs" described, 231
dragons/dragon-fighting analogy, 111–112
drip torch described, 22
Duc, Thich Quang, 84, 85
Dunham, Brandon, 112, 113
Dunn, Cassie, 129, 130, 133
Dust Bowl, 25
Dwight, Timothy, 69

ecological benefits of fire
    fire crew beliefs, 39–40
    plants-animals relationships, 60
    prairie, 10–11, 24–25, 55–56
    restoration, 37, 85–86, 105–107, 216–217
    Rim Fire, 188–189
    Yosemite National Park/giant sequoias, 190, 210–212
    See also fires, prescribed; *specific fires/locations*
Ecological Restoration Institute, Northern Arizona University, 222
"ecology of attention," 188
Ecology of Plant-Derived Smoke, The, 75
Ecosphere (journal), 282
ecosystem management vs. species management, 266–267

Index

Edwards, Dianne
  background/career, 42
  fire/ancient plants, 42–43
  lightning/fires, 42, 44
Effects of Burning of the Grasslands and Forest by Aborigines the World Over/Forgotten Fires (Stewart), 56, 57
Egan, Jack, 3
Eldorado Hotshots, 128
Ember Alliance, 201
Endangered Species Act, 217, 266–267
Energy Release Component, 125
Enlightenment and "monsters," 112
Ensenada RX prescribed burn (Carson National Forest)
  archeological sites, 232–233, 245–246
  area descriptions, 231
  crew, 231
  descriptions/fire work, 231–241, 243–245
  FEMO, 232, 235–237, 238–239
  helicopter/dragon eggs, 231, 232, 239
  objective/ecological benefits, 232
  O'Connor, 229, 230–241, 243–246
  safety, 232, 233
environmental groups-forest management controversy, 208–209, 217–219, 220, 221, 319
Eospermatopteris, 46
eucalyptus/Eucalyptus trees, 4–5, 6
eugenics/movement, 89, 90, 94
extreme fires defined, 100
  See also megafires

Faulkner, William, 248
FEMO (fire effects monitor) duties, 235
Ferguson Fire/firefighters' deaths, 193, 197
Fire Adapted Network, 202
"fire bug," 63, 64
fire effects monitor (FEMO) duties, 235
"fire festivals" of South, 247–248
firefighters
  bonds/camaraderie, 127–128, 193
  communications significance, 236
  crew member death, California, 317
  fire relationship, 76–79
  heat/smoke effects, 36–37
  morning briefings/evening debriefings, 14
  Nomex uniform described, 15
  respiratory issue/"camp-crud," 173, 188
  suicides and, 130–131
  See also specific fires/components; specific individuals
Fire in America (Pyne), 204
"fire mistresses," 83–84
fire modules, 197–198
Fire on the Rim (Pyne), 191, 192
fire pack basics, 21
fire paradox, 8
fire revolution, America
  beginnings/descriptions, 198–200
  Yellowstone fires (1988) and effects, 200
fires
  Black woman/Forest Service official incident, 254
  chains per hour, 118
  descriptions, 5, 24, 31–32, 41, 55, 61–62, 111–112
  dragon descriptions, 111–112
  grass vs. forest fires, 21
  ingredients needed, 23–24, 43
  kinds of (Pyne), 185
  moving uphill vs. downhill, 118
  nighttime, 142–143
  plants dependent on (overview), 4–6
  regulations beginnings, 70
  settlers in America, 51, 88, 261–262
fire scars
  Blodgett Experimental Forest, 105
  dendrochronologists, 58–59
  formation described, 57
  Hecate Island, British Colombia, 59
  Jemez Mountains, New Mexico, 58–59
  Laboratory of Tree-Ring Research, Arizona, 58
  temperate rainforests, 59

fire scars (*continued*)
   tree-ring fire-scar network, North America, 58–59
   tree sampling and, 57–58, 59
fires/early Earth
   ancestors/consequences, 47–49
   ancient plants, 5, 42–43, 46–47
   first fires, 42–44
   hearths, 47–48
   plants/ingredients, 44
   See also specific components/locations
fires, prescribed
   anti-fire forces and, 251–252, 253–254
   "army of fire practitioners," 201–202
   Bootleg Fire and, 141
   burn windows, 298
   changing human perspective, 243
   climate change, 107, 193, 298, 316–317
   descriptions, 19–20
   early southern states, 284–285
   fire deficit, 233, 243
   fire revolution, America, 198–201, 214–215
   forest openings/density, 217
   Forest Service/escaped fires and, 241–243
   Indigenous people/BIA and, 305–306
   Minnesota/The Nature Conservancy (1962), 198
   moratorium/background, 316–317
   movement locations summary, 10
   precolonization state/parks, 215
   prescribed-burn associations, 26–27, 39
   Pyne on, 187, 190, 192–193, 198–199, 200–201, 205–206
   Rim Fire, 188–189
   terms for, 10, 205
   timidity and, 263
   Washburn Fire, 318–319
   Yosemite National Park/giant sequoias, 194, 210–213
   See also ecological benefits of fire; specific components/locations; specific individuals/organizations

fire suppression
   antifire campaigns, 90–91, 92–94
   basic tactics, 142
   beginnings/concept spread, 7–8, 87–88
   budget/BIA, 305–306
   class, 88, 103
   consequences/legacy, 7–8, 91–92, 96–98, 103–104, 189, 214–215, 216, 217
   data collection, 103
   dog use, 93–94
   European attitudes, 87–88
   following Yellowstone fires, 200
   Klamath region, 280, 281–282, 283, 284–285
   light-burners, 91–92, 94
   Martin on, 303, 304, 305
   megafires, 103–104
   nighttime, 142
   paradigm shift from, 217
   psychologists (1930s/1940s), 92–93
   punishment for fires, 62, 90–91, 93–94, 103, 252, 281, 286–287
   Southern Forestry Education Project, 92
   US agencies/individuals (nineteenth/early twentieth century), 89–97
   warnings on, 91–92, 106–107
   West on, 133–134
   World War II and, 94–95
   Yellowstone fires/effects, 200
   Yosemite/effects, 189
   See also specific individuals/locations
fire triangle, 23–24
Fire Victim Trust, 180
fire whirl, 122–123, 244
First World Congress on National Parks (1962), 198
Flagstaff area
   "Big One" and, 221
   WUI, 221
   See also restoration project/Flagstaff area
Flagstaff Hotshots, 117, 118–119, 120, 121, 126

Index

flanking fires described, 19
flapper description/use, 20–21, 255
Flematti, Gavin, 74–75
Forest Guardians, 95
forest management controversy, 208–209, 211–212, 217–219, 220, 221, 319
Forest Service, US, 16, 17, 28, 64, 69–70, 89, 91, 92, 94, 95, 103, 110, 113, 133, 142, 217–218, 224, 231, 232–233, 251–252, 254, 279, 299–300, 316–317
Forest Stewards Guild, All Hands All Lands, 233, 234
Forgotten Fires/Effects of Burning of the Grasslands and Forest by Aborigines the World Over (Stewart), 56, 57
Four Forest Restoration Initiative, 223–224
Fremont-Winema National Forest, Oregon, 140
French Revolution, 83
Frierson, Cathy, 83–84
Fromm, Michael
  position/career, 99
  pyroCbs, 99, 100, 111
FUSEE: Firefighters United for Safety, Ethics, and Ecology, 104

Gagadju people, 4
Gammage, Bill, 2, 3, 4
Garry oak ecosystem, Willamette Valley, Oregon, 60
General Sherman/fire wrap, 209–210
Gentiana catesbaei, 4–5
Geological Survey, US, 215
Giant Forest Grove, 210, 212–213
Giant Sequoia National Monument, 215
giant sequoias/groves
  chainsaw bar sizes, 213
  climate change, 215–216
  conservation movement, 207
  debate on protection/fire, 207–210, 212–215, 319
  descriptions/distinctiveness, 203, 207, 211–212
  fire deficit, 207–208, 214
  fire mortality, 207, 210, 215–216
  General Sherman/fire wrap, 209–210
  Grizzly Giant, 203, 318
  KNP Complex/opportunities, 210–211
  Mono wind, 202–203
  prescribed burns, 199, 210–214, 215
  thinning trees/controversy, 208–209, 210–211, 217–219
Giant Sequoias of the Sierra Nevada, The, 199
Gifford, Neil, 72
Gila National Forest, 200, 212
Gilboa, New York primordial forest, 45–47
Gimlin, Bob, 288
Glass Fire (2020), 296
Gleason, Paul, 77
global warming. *See* climate change
Goldring, Winifred, 46
Grand Canyon National Park
  Native tribes/evidence and fire, 226–227
  Native tribes removal, 225–226
  Nez on, 225–228
Granite Mountain Hotshots/deaths, 130
Grassroots Wildland Firefighters, 304
Great Depression, 52, 81, 251
Great Oxidation Event, 43–44
Green Swamp, North Carolina, 269–272
"green, the," 21, 119
Grizzly Giant, 203, 318
Guide to Nabokov's Butterflies and Moths, A (Zimmer), 70

Haines Index, 118
handline described, 18–19
Hanh, Thich Nhat, 85
Happy Camp Complex, California (2014), 112–113
Haraway, Donna, 7
Harling, Will, 278, 315, 320
Harper, Roland, 249, 250, 253
Hartesveldt, Richard, 199
Havasupai, 226
head fires described, 19
Hermits Peak/Calf Canyon Fire, New Mexico (2022), 316–317

Hiawatha National Forest, 16
Hollandophyton colliculum, 42–43
Holloway Fire, Nevada-Oregon state line (2012), 124
Homo erectus, 47
Homo naledi, 47
Homo sapiens, 47
Hostler, Clarence, 307–308
hotshot crews
  beginnings, 95
  descriptions, 18, 115–116, 234
  See also specific crews/individuals
Hotshot Wake Up podcast, 145
House Committee on Oversight and Government Reform, US, 304
Howarth, Coby
  background/fire experience, 155–157, 159, 177
  description/traits, 155–157, 172
  Dixie Fire/crew, 155, 156, 159, 172, 177
  "smashing," 156, 157
  wife, 155–156
Huffman, Dave, 222
Hughes, Brian, 197
Humphreys Peak, Arizona, 219–220
Humphries, Aaron, 127–128
Hupa people, 273, 278–279, 285
  See also Klamath region
Huron people, 75
Hurricane Irene, 46

Ice Age/Pleistocene epoch and charcoal record, 49
ice capping, 120
ignition of fires, earliest
  asteroids, 44–45
  lightning strikes, 42, 44
  volcanoes, 44
Illilouette Creek Basin, Yosemite National Park, 189–190
Incident Response Pocket Guide (National Wildfire Coordinating Group), 156
InciWeb, 65

Indigenous people
  Aboriginals/fire, 2, 4
  cultural trees, 240
  European attitudes towards, 259
  fire, 2, 4, 10
  sovereignty, 10, 278–279
Indigenous people, Native Americans
  abortifacient (peeled cambium), 240
  in California/European settlement time, 60–61
  cultural trees, 240
  environmental activists, 218–219
  genocide of/landscape consequences, 7
  Grand Canyon, 225–227
  Humphreys Peak/San Francisco Peaks, Arizona, 219–220
  national parks authority and, 305
  peyote use, 52–53
  sovereignty, 10, 278–279
  See also specific peoples/individuals
Indigenous people, Native Americans/fires
  American West, 10, 88–89, 90, 244, 307–308
  anthropogenic ecosystems/examples, 59–61
  burned landscapes descriptions, 54–55, 59–60, 61
  California, 60–62
  Catskills tree species, 50–51
  imprisonment and, 62
  Jemez Mountains, 219
  knowledge/art of, 261–262
  Meriwether Lewis on, 244
  pine barrens, 65, 69, 72
  prairies, 6, 24–25, 55–56
  prescribed burns, 305–306
  purposes, 6, 24–25
  in South, 253, 261
  Stewart's evidence, 53–54, 55–56, 57
  "tending/cleaning the land," 53, 61
  See also specific peoples/individuals
Ingalsbee, Timothy, 104
Initial Attack crew, 144

Index

Intergovernmental Panel on Climate Change reports/causes, 215
International Association for Wildland Fire, 248
International Journal of Wildland Fire, 104
internment camp, Japanese, World War II, 180, 181

Jemez Mountains, New Mexico fire history, 58–59
Jemez people
　depopulation/landscape consequences, 59
　fire, 58–59
John Muir Project, California, 217–218
Jump Dance, 274, 276, 300

Kaibab National Forest, Arizona, 223, 225, 227
Kansas's Flint Hills fire, 200
Karner blue butterflies
　decline, 68–69, 70–71
　discoveries, 67–68
　fire, 70–71, 75–76
　increase, 75–76
　mutualistic relationships, 69
　See also Albany Pine Bush
Karner, New York and Lepidoptera, 66–67
karrikinolides, 74–75
Karuk people, 61, 273, 278–280, 281, 283, 285, 290, 299, 303, 307
　See also Klamath region; Yurok people
Karuk Tribe Department of Natural Resources, 61, 303
Kelleher, Dan
　background/fire experience, 15, 17–18, 20, 22–23, 24, 29–31, 34, 35, 36, 38, 39–40, 77, 140–141, 143, 144, 145, 188, 315, 321, 322, 323
　description/traits, 15–16, 17, 302–303
　Dixie Fire/area, 139–140, 144, 145, 181
Kestrel meter, 235
King, Martin Luther, Jr., 85

Kiowa Mountain, 231
Klamath Mountains, 61
Klamath Mountains, The: A Natural History, 283
Klamath National Forest, 91, 278–279
Klamath region
　climate change, 282–283
　dams/fish, 280
　fire suppression, 278, 280, 281–282, 283, 284–285
　Forest Service planned highway/spray, 279–280, 308
　Gold Rush, 308
　"government-subsidized capitalism," 278–279
　logging, 278, 279, 281, 290, 308
　Native land, 278–279
　PTSD in locals/fire suppression, 278
　thinning needs, 290
　wildfire/pyroCbs, 277–278
　wildland firefighters' views, 277–278
　See also Yurok people
Klamath region prescribed burn
　Cultural Fire Management Council, 285, 286, 287
　descriptions/fire work, 283, 287, 293–298, 301–303, 307–310
　Hee' Mehl unit description, 293, 294
　O'Connor, 283, 287–288, 293–298, 301–303, 307–310
　Raven, 296–297, 308
　women, 302–303
Klamath River Jack, 91
Klamath Tribes, Oregon, 285
　See also Klamath region; *specific tribes*
KNP Complex, California (2021), 209–211, 263
Kolden, Crystal
　career, 101
　fire/management, 101–103, 218
Komarek, Betty, 262
Komarek, Ed, 254, 259, 261–262
Kosek, Jake, 94–95, 231, 242
Kroeber, A. L., 52, 53, 56, 276
Kudish, Michael, 50–51

Laboratory of Tree-Ring Research, Arizona, 58
Lake, Frank Kanawha
  background, 299–300
  fire/benefits, 299–302, 306
  spirituality of, 299–302
Langstroth Fire, 212
Lassen Hotshots, 114, 115, 117
Lassen National Forest, 134
"LCES," 77, 131–132, 257
Leopold, Aldo, 198–199, 220, 248
Leopold, Starker/report, 198–199, 214–215, 216
Lepidoptera, Karner, New York, 66–67
Lessons Learned Center
  purpose, 131–132
  Taylor, 131–132
  West, 131–132
Levy, Sharon, 290
Lewis, Meriwether, 244
light detections/ranging (lidar) signals, 98–99
  See also pyroCbs
lightning strikes/earliest fire ignition, 42, 44
Loma Prieta Earthquake, California (1989), 142
Long, Ellen Call, 253
longleaf pine ecosystem/fire, 92, 248, 249–250, 251–252, 253, 256, 258, 260–261, 264–266, 267–268, 269–272
Longshots, 191
Lookout, The blog, 142, 143, 173, 186–187
Lopez, Gene
  background/fire experience, 144–145, 148, 158, 159–163, 165, 166–167, 174–177, 178
  Dixie Fire/crew boss, 144, 148, 158, 159–163, 167–168, 174–177, 178
  Greenville, 160–163
  on job, 165
  leadership style, 159–161
Lopez, Valentin, 10
Los Alamos National Laboratory/ radioactive material, 242–243

Los Angeles Times, 218
Ludlow Bone Bed, 42
Lukens Fire, Yosemite National Park, 195, 197, 206
Lunder, Zeke
  background/fire experience, 35, 37–39, 66, 141–143, 173, 315
  Dixie Fire analysis, 141–143, 173, 186–187
  Lookout, The blog, 142, 143, 173, 186–187
  Marin County, California, 38–39
Lycaeides melissa samuelis, 67
Lycaena scuderii, 67
Lycopsida, 47

MacArthur Fellowship/"genius grant," 192
McConnell, Bob
  on climate change, 283
  on dams/fish kills, 280
  description/traits, 273
  fires/fire suppression, 282-yur283, 284–285, 287
  Yurok dances/ceremonies, 274–275, 276, 277
McCovey, Kathy, 290
McCovey, Mavis, 279–280, 284
Maclean, Norman
  background/fire experience, 110
  Mann Gulch wildfire, Montana (1949), 109–111, 183
  on smokejumpers, 109
McRae, Rick, 100
Mann Gulch wildfire, Montana (1949)
  descriptions, 21–22, 109–111
  fire descriptions, 111
  pyroCbs, 111
  Young Men and Fire (Maclean), 109
Mariano, Chris, 128
Martin, Kelly
  background/fire experience, 303–305
  as whistleblower/women's treatment, 304
Maxwell, Hu, 54–55

Index

Medicine Trails (McCovey), 284
Medio Fire, New Mexico (2020), 306
megafires
   climate change, 8, 101, 104–105, 107
   communications, 101–102, 103–104
   descriptions, 100–101
   fire suppression and, 103–104
   predictions, 102
   term use beginnings, 100
   tree mortality with, 106
Megram Fire, California (1999), 299–300
mental health/firefighters
   Lessons Learned Center, 131–132
   stigma, 128
   See also PTSD
Merced Grove, 208
"missing fire," 233
Modoc National Forest, 129–130
Modoc Tribe, 296
Mono wind, 202
Montyel, Marandon de, 80
mop-up (cold-trailing) described, 152
Morton, Thomas, 54
"Moses Letter," 64
Mueller, Jennifer, 201–202
Muir, John
   background/prairies, 90, 95
   conservation movement, 89–90, 95
   fire suppression views, 89–90
   racism/eugenics movement, 90
   religion, 90
   Yosemite National Park, 89–90, 190, 203
Murie, James Rolfe (Sa-Ku-Ru-Ta), 34
Museum of Comparative Zoology, Harvard, 67
Musser, Paul, 118–119, 121
mycorrhizae, 68–69
Myers, Frankie, 280, 281, 310

Nabokov, Vladimir
   background, 67
   butterflies/moths, 67–68, 70
Nachusa Grasslands, 220
Nagorno-Karabakh War, 150

Napa Valley and fires, 135, 296
National Park Service, 193, 242
National Wildfire Coordinating Group, 156
Native American Church, 52–53
Native Americans. *See* Indigenous people, Native Americans
Natural History of the Albany Pine Bush (Barnes), 69
Nature, 102
Nature Conservancy's Indigenous Peoples Burning Network, The, 285
Nature Conservancy, The, 63, 64, 198, 202, 285
Navajo (Diné people), 225, 226, 228, 240, 321
Neanderthals, 47, 48
Neel, Leon
   fire/fire need, 260–261, 263
   old-growth trees, 260
   Stoddard and, 260
   on wilderness preservation history, 260
Neidjie, Bill, 4
New Mexico's pueblos, 285
News and Observer (newspaper), 82–83
Newsom, Gavin, 168
New Zealand and charcoal record, 49
Nez, Jason
   Grand Canyon National Park/career, 225–228
   Native tribes/people, 225–228, 321–322
Nomex material, 15
Norgaard, Kari Marie, 90–91, 281
Nussbaum, Martha, 6–7
Nuttall Complex, Arizona
   descriptions/fire work, 116–123
   firefighter with rhabdomyolisis, 122
   fire shelters, 123, 126
   H4 safety zone, 121, 122–123, 126
   pyroCb, 120
   Taylor, 121, 122–123, 126, 131
   West, 116–120, 131

oak bottleneck, 51
oaks and fire, 51

O'Connor, Eric, 225, 312, 313
O'Connor, M. R.
　conflict over fire work, 229–230
　FEMO coursework, 235
　firefighter's illness/injuries, 173, 188
　firefighter's predilection and, 207–208
　miscarriage/emotional consequences, 37
　search for crew work/following Nebraska, 64–66
　son/rhinovirus, 229, 230
　wildland-firefighting classes/tests, 8–9, 14, 23–24, 235, 311–314
　See also specific locations/activities
O'Connor, Sandra Day, 279
Oglala Lakota, 314
Ogle, Charles, 91–92
Ojibwa, Bay Mills Indian Community, 16
Ojibwe, Leech Lake Band, Minnesota, 285
Okanogan-Wenatchee National Forest, Washington, 124
On the Fireline (Desmond), 78
Orcasitas, Bre, 77–78, 104, 143–144
O'Rourke, Thomas, 277
Osborn, Henry Fairfield, 90
owl, spotted/protection and consequences, 95, 223
oxygen
　cyanobacteria, 43–44
　Great Oxidation Event, 43–44
　as requirement for fire, 23–24, 43

Pacific Crest Trail, 195
Padillo, Alfredo, priest, 231, 313
"Paiute forestry," 92
Paleolithic (Upper) period and fire, 49
Paleolithic people, 245–246
Pamlico Sound, North Carolina, 263
Paradise, California
　firefighters' search/human remains, 165–166
　Frank (wildland firefighter), 179–180, 181–182
　See also Camp Fire/deaths, California (2018)

Pathological Fire-setting (psychiatrists/Columbia University), 79–80
Patterson, Roger, 288
Pawnee, 24–25, 26, 34
Payette National Forest, 317
Peterson, Melissa, 128
PG&E, 135–136, 179–180
Pier Fire, California, 216
Pinchot, Gifford
　eugenics movement, 89
　forestry/fire, 89
pine barrens, New Jersey/New York
　butterflies-lupine relationship, 67
　description, 65
　European settlers, 69
　fire needs, 68–69
　fire regulations, 70
　Lepidoptera/decline, 66–68, 70–71
　Native Americans/fire, 65, 69, 72
　See also Albany Pine Bush/fire
Pine Ridge Reservation, South Dakota, 314
Pioneer Fire, Boise National Forest
　"ghost" wildland firefighter, 166–167, 175–176
　Lopez, 166–167, 175–176
Pleistocene epoch/Ice Age and charcoal record, 49
Plumas Hotshots, 117, 119
Plumas National Forest/fires (California)
　Beckwourth Complex, 144
　Dixie Fire, 146, 147, 151
　engine malfunction/burning incident, 130
Pnin (Nabokov), 68
pocosin description/fires, 256–257
Pomo Indian tribes, California, 53
ponderosa pine forests and fire, 64
PPE (personal protective equipment), 15
prairie
　climate and fire, 55–56
　descriptions, 22, 25
　ecological benefits of fire, 24–25, 55–56
　grass types and fire, 23

Index

grass vs. forest fires, 21
trees/eastern redcedars, 25
prairie chickens, 26, 28
prairie prescribed fire, Ord, Nebraska
  activities mornings/evenings, 27–28
  blacklining, 30–31
  crew/emissaries of good fire, 34–35
  description of area, 13
  fire crew incident command post, 13–14
  O'Connor, 10–11, 13, 15, 17–18, 22, 23–24, 28, 29, 30, 31–32, 33–34, 35, 36–38, 39–40
  purpose, 11
  wildland firefighters descriptions, 28–29, 31
  See also specific individuals
prairie prescribed fire, Ord, Nebraska second crew
  background, 315
  blacklining, 322–324
  burning 210-acre unit, 319–324
  burn permits' cancellations, 316
  community trust and, 315, 320
  drought/wind, 315–316
  O'Connor, 315–316, 319–324
  preparation, 315–316
  weather change/rain, 324
prairyerth described, 24
Prescribed Burning Act (Florida/1990), 285
prison camp/fire-training, 147, 150–151, 160, 180, 317
Proceedings of the National Academy of Sciences, 48
Promontory Cave, Utah, 52
PTSD
  Diagnostic and Statistical Manual of Mental Disorders criteria, 128–129
  suicides and, 130–131
  symptoms, 113
  Taylor, 126–127, 160
  West, 113, 132, 133, 135
  Yurok people, 278
Pueblo of Tesuque, 306

Pulaski tool, 117, 148, 149, 255
pulling hose described, 172
Pyne, Stephen
  background/fire experience, 57, 87, 188, 191–192, 193–194, 206, 227
  Buckley and, 194–198, 203
  description/traits, 189
  fire/history of fire, 57, 87–89, 90, 92, 93, 95, 185–187, 190–192, 196, 198–199, 203–206, 247, 249
  on intentional/prescribed fire need, 187, 190, 192–193, 198–199, 200–201, 204–206
  kinds of fire, 185
  Longshots, 191–192
  mother, 193
  O'Connor/Yosemite, 188–189, 193–196, 201–206, 207–208, 318, 319
  on South/fires, 247, 249
  Yosemite Park hypothesis, 189–190
  Yosemite Park research/fire, 188–190, 193–196, 201–206
pyroCbs
  Bootleg Fire, Oregon, 140–141
  climate change, 112
  descriptions, 99, 111–112
  downdrafts, 111–112
  fire tornadoes, 112
  light detections/ranging (lidar) signals, 98–99
  location detections, 99–100
  updrafts, 140–141
  Yarnell Hill Fire, 130
  See also specific fires
"Pyrocene," 185–186, 193
Pyrocene, The: How We Created an Age of Fire and What Happens Next (Pyne), 190, 193
pyromaniacs/women
  background/abuse, 80–81, 82
  descriptions/examples, 80–84
  reasons (historical theories), 80–81
  Russia, 83–84
  Samarcand Manor, 81–83

pyromania/pyromaniacs
  developmental times, 79–80
  historical fears/actions, 83–84
  medical diagnosis beginnings, 83
  mid-twentieth century descriptions/
    examples, 79–80
  nineteenth century descriptions/
    examples, 80–81
  Russia, 83–84
pyrophyte described, 4, 50

Qesem Cave, Israel, 47

racism, 89, 90, 94–95, 254
Railroad Fire, California, 216
Ramsey, Jayson
  background/fire experience, 146, 147,
    149, 157, 158, 172, 176, 178, 182–183,
    186
  description/traits, 157–158, 172, 187
  Dixie Fire/crew, 146, 149, 157, 158,
    172, 176, 178, 182–183, 186
Rangel, Garret
  background/fire experience, 146, 147,
    150, 151, 152, 157, 158, 160–161,
    172, 187
  Dixie Fire/crew, 146, 147, 150, 151,
    152, 158, 160–161, 172, 187
Raven Parking, 296
Reagan years, 199–200
Red Earth, White Lies (Deloria), 52
Redwood Mountain, 199
Redwood National Park, 278–279
Reserved Treaty Rights Land Program,
  306
restoration
  concept, 222
  controversy overview, 216–219, 223
  Four Forest Restoration Initiative,
    223–224
restoration project/Flagstaff area
  environmentalists' criticism, 221
  forest density/openings statistics, 222
  forests of past/knowledge, 221–224
  goal/precolonization state, 220

logging in past, 222, 223
money sources, 221
O'Connor, 219, 222–223, 224–228
prescribed burn plan, 223
silviculturalist, 223–224
thinning trees, 220, 221, 224–225
TREX event, 220
WUI/conditions, 221
See also Flagstaff area
rhino tool, 149, 169
Rim Fire (2013), 188–189
Roadless Wilderness Area designation,
  279
Robbins, Margo
  description, 288
  fire, 284–285, 291–292, 294, 302–303,
    309, 310
  fires/basketry, 283, 284, 288–289,
    290, 291
  plant knowledge, 288–289
Rough Fire, California, 215–216

Sacramento Bee, 218
safety and fires, 14, 21–22, 35–36, 48–59,
  77, 131–132
  See also specific regulations/fires
Sagan, Carl, 98–99, 100
Sagard, Gabriel, 75
Sa-Ku-Ru-Ta (Murie, James Rolfe), 34
Salmon and Acorns Feed Our People
  (Norgaard), 90–91
Samarcand Manor/pyromaniacs, 81–83
San Carlos Apache Reservation,
  Arizona, 205
Sand Count Almanac, A (Aldo Leopold),
  220
San Francisco Peaks, 219–220
sawyer position/work, 127
Schafale, Mike, 256, 257
Science, 214
Science of the Total Environment, 298
Scott Fire, Grand Canyon (2016), 227
Scudder, Samuel H.
  background/butterflies, 66–67
  Scudder's blue/Karner blue, 67

Index

seed germination and butenolides/karrikinolides, 74–75
self-immolations
　examples, 84–85
　reasons, 84–85
Sequoia and Kings Canyon National Park, California, 193–194, 197, 199, 209, 210
Sequoia National Forest, 218
sequoia trees. *See* giant sequoias/groves
Shanidar Cave, Iraq, 48
Shea, John P., 92–93
Sheehy, Luke, 129–130
Sheep Fire, 133
shelters, fire shelters
　American wildland firefighters, 123
　descriptions/use, 15, 21, 123–124
　Granite Mountain Hotshots/deaths, 130
　qualifications test, 123–124
Shoshone, 52, 309
Sierra Club, 95
Silcox, Gus, 94
silk cultivation, 253
silviculturalist described, 223
Sioux, 311
Six Rivers National Forest, 278–279
Skunk Fire, Arizona (2014), 205
Sky Islands, Arizona, 116
Sleeping Bear Dunes National Lakeshore, Michigan, 65–66
sling psychrometer described/use, 235–236
smoke
　agricultural use, 75
　components, 74–75
smoke inversion description, 148
smokejumpers, Mann Gulch fire, 109
smoke water, 75
Smokey Bear, 4, 87, 94, 95, 96, 195, 248
snow line-wolf lichen, 173
South and fire
　as art, 249, 260, 261–262
　logging, 249–250, 259–260
　longleaf pine ecosystem, 92, 248, 249–250, 251–252, 253, 256, 258, 260–261, 264–266, 267–268, 269–272
　overview, 247–249
　plantations/Blacks, 253, 254
　See also specific individuals/locations; Tall Timbers; Wilmington, North Carolina area/prescribed burns
Southern Forestry Education Project, 92
Southwest fire deficit, 233
Spanish in North America, 59
species management vs. ecosystem management, 266–267
Spott, Robert, 281
Standing Ground (Buckley), 279
stand-replacing fire, 96
Stanislaus National Forest, California, 193
Stein, William
　background/career, 45
　primordial forest/Gilboa, New York, 45–47
Stephenson, Nate
　background/fire experience, 215–216
　climate change, 215–216
　on forest management/controversy, 219
Stephens, Scott
　career, 105, 190
　fire/megafires, 105–106, 107, 190
sterilization of girls (1929-1950), 82
Stewart, Omer
　background/religion, 52–53, 276
　book on fire, 56, 57
　fire/intentional burning, 53–54, 55–56, 57, 259–260
　Tall Timbers conference, 259–260
Stoddard, Herbert
　background/youth, 250–251
　Black woman/Forest Service official incident, 254
　bobwhite quail, 251–252
　fire/fire need, 250–252, 254, 261, 263
　old-growth trees, 260
Stowe, Johnny, 248

"student of fire," 77
suicides
    wildland firefighters, 130–131
    See also self-immolations
Supreme Court, US, 279
Susanville, California and fires, 135, 136, 159
Suwanee Grove, 210
"swampers," 127
Swetnam, Tom
    career, 58, 95, 98, 213, 217, 218, 219
    Cebollita Fire, 97, 98
    father, 95–96, 97–98
    forest management controversy and, 218, 219
    giant sequoias/fire, 212–214, 215, 218
    Jemez Mountains/fire, 58–59, 95–96, 97, 98
Sycan Marsh Preserve, 140, 141

Tall Timbers
    cofounders, 253, 254
    Fire Ecology Conference (1962), 198
    first conference, 253
    origins/purpose, 253, 260
    second conference (1963), 259–260
    seventh conference/religious-like fervor, 262–263
    Stoddard/Beadel, 253
Talth people/prophecy, 282
Taylor, Thomas ("Thom")
    background/fire experience, 120–121, 122–123, 124–127, 131, 317–318
    fire shelters, 123, 124–126
    mental issues/PTSD, 126–127, 131–132, 160
teak forests, 88
"10 a.m. policy", US Forest Service, 94
Tending Fire (Pyne), 88
Tending the Wild (Anderson), 56–57, 216
thinning trees/controversy
    description, 216, 217–218, 219, 220, 221, 224–225, 231, 290
    giant sequoias/groves, 208–209, 210–211, 217–219

Thirtymile Fire, Washington
    description/fire work, 125–126
    fire shelters/deaths, 125–126
    pyroCb, 125
Thompson, Lucy, 275–276, 282
Thoreau, Henry David and accidental fire, 76
Tonto National Forest, Arizona, 223
tornadoes from fire, 97, 99, 102, 112
To the American Indian (Thompson), 275–276
training exchange (TREX)
    creation of, 63–64
    described, 14
Treuer, David, 305
TREX. *See* training exchange (TREX)
Tripp, Bill, 61
Tsing, Anna, 112
Tunguska asteroid/consequences, 44–45
Tuolumne Grove, 208

Umpqua National Forest, Oregon, 115
Understories; The Political Life of Forests in Northern New Mexico (Kosek), 94, 231
updrafts and pyroCbs, 140–141
Upper Paleolithic period and fire, 49
urban sprawl and fire, 200
urban wildfire zone, 221
US Cavalry, 89
Ute, 240

Valley Fire, California (2015), 39
van Wagtendonk, Jan, 200–201, 213, 318, 319
van Wagtendonk, Kent, 318
Vietnam War/protest, 84
volcanoes/early ignition of fires, 44

Wagon Burners, 296
Waltz, Amy, 221
War Advertising Council, 94
Washburn Fire, California, 318–319
Washoe Tribe, California/Nevada, 285
wedge prisms description/use, 223

Index

West, Mike
  on *Anchor Point, The*, 112–113
  background/fire experience, 112–120, 127, 160
  on climate change/fire, 134–135
  description/traits, 114, 115
  father, 114
  on fire suppression, 133–134
  Happy Camp Complex, California (2014), 112–113
  mental issues/PTSD, 112–113, 114, 127, 129, 130–133, 135, 160
  Nuttall Complex, 116–120
  on safety, 134–135
  sawyer recommendation/O'Connor, 311–312, 313
  Sheehy/Sheehy's death, 129–130
  Susanville and fire, 133, 135, 137, 147
  therapy/new career, 133
West, US
  fire regimes, 89
  overgrazing, 88–89
  See also specific locations/fires
Wheeler, Ben
  background/fire experience, 26–27, 34–35, 76–77, 137, 315, 316, 319, 320–321
  childhood/fire, 76–77
  son, 321
  vision/work, 26–27
White Deerskin Dance, Yurok, 274–275, 280, 282, 300–301
Whyte, Kyle Powys, 307
Wildcat Fire, Grand Canyon, 227
wilderness concept/myth, 6–7, 52, 90, 218–219, 226, 259–260
Wildfire Magazine, 100
wildland firefighters. *See* firefighters
Wildland Fire Mitigation and Management Commission, 303
wildland-urban interface (WUI), 221
Wildlife Resources Commission, North Carolina, 263
Willamette Valley, Oregon, Garry oak ecosystem, 60

Wilmington, North Carolina area/prescribed burns
  black rails/habitat, 266–267
  descriptions/fire work, 249, 254–259, 263–272
  Drew, 254–255, 264, 268, 269, 270
  Green Swamp/escaped fire, 269–272
  landscape descriptions/diversity, 255–257, 270–271
  leaf blower/use, 255
  Marsh Master, 264–266
  O'Connor, 249, 254–259, 263–272
  safety, 257, 266
  Travis, 264, 265, 266–269
Wind Cave, South Dakota, 311
Windy Fire, California (2021), 210, 212
Wonderwerk cave, South Africa, 47
Work Capacity Test, 14
World Fire: The Culture of Fire on Earth (Pyne), 89
World Meteorological Organization, 99
Wounded Knee cemetery, South Dakota, 314

"yards"/"fire yards," 60
Yarnell Hill Fire/deaths, Arizona, 130, 236
Yellowstone National Park
  establishment, 89
  fires (1988) and effects, 200
Yosemite National Park
  "Firefall," 194, 206
  Illilouette Creek Basin, 189–190
  Muir, 89–90, 190, 203
  origins, 189
  prescribed burns/letting fires burn, 189–190, 194, 200–201
  Pyne/research, 188–190, 193–196, 201–206, 207–208, 318, 319
  See also giant sequoias/groves
Young Men and Fire (Maclean), 109
Yurok people
  acorns, 287–288, 289–290
  on arrival of white people, 307
  arson/arsonists, 285–287

Yurok people (*continued*)
　basketry/baby basket, 283–284, 289–290
　beliefs/values, 273, 274, 274–277, 282, 283–284, 291, 294, 307–308
　carbon sequestration, 306
　creation story, 273
　Cultural Fire Management Council, 285, 286, 287
　dances/ceremonies, 274–277, 280, 282, 300–301
　fire creation story, 291
　fires/benefits, 61–62, 275, 280–281, 283, 284, 285–287, 288–292, 294, 300–302, 306, 307–308, 310
　fire suppression and, 278, 280–283, 283–287, 289, 290, 299–300, 301
　lack of sovereignty/fire sovereignty, 278–283, 301–302, 303, 305
　punishment for fires, 62, 281, 286–387
　regalia, 275–277, 300–301
　stick game, 289
　wrestling/mixed martial arts, 289
　Yurok Reservation location, 274
　See also Klamath region; *specific individuals*

Zimmer, Dieter, 70
Zipf, Karin L., 81–82
Zuni Hotshot, 124
Zuni Mountains, 234
Zuni people, 240

Credit: Stephen Pyne

M. R. O'Connor's reporting has appeared in the *New Yorker, Foreign Policy, The Atlantic, Nautilus, UnDark,* and *Harper's.* She is a graduate of Columbia University's Graduate School of Journalism and was a 2017 Knight Science Journalism Fellow at MIT. She lives in Brooklyn with her partner, the screenwriter Bryan Parker, and their two sons.

PublicAffairs is a publishing house founded in 1997. It is a tribute to the standards, values, and flair of three persons who have served as mentors to countless reporters, writers, editors, and book people of all kinds, including me.

I. F. STONE, proprietor of *I. F. Stone's Weekly*, combined a commitment to the First Amendment with entrepreneurial zeal and reporting skill and became one of the great independent journalists in American history. At the age of eighty, Izzy published *The Trial of Socrates*, which was a national bestseller. He wrote the book after he taught himself ancient Greek.

BENJAMIN C. BRADLEE was for nearly thirty years the charismatic editorial leader of *The Washington Post*. It was Ben who gave the *Post* the range and courage to pursue such historic issues as Watergate. He supported his reporters with a tenacity that made them fearless and it is no accident that so many became authors of influential, best-selling books.

ROBERT L. BERNSTEIN, the chief executive of Random House for more than a quarter century, guided one of the nation's premier publishing houses. Bob was personally responsible for many books of political dissent and argument that challenged tyranny around the globe. He is also the founder and longtime chair of Human Rights Watch, one of the most respected human rights organizations in the world.

• • •

For fifty years, the banner of Public Affairs Press was carried by its owner Morris B. Schnapper, who published Gandhi, Nasser, Toynbee, Truman, and about 1,500 other authors. In 1983, Schnapper was described by *The Washington Post* as "a redoubtable gadfly." His legacy will endure in the books to come.

Peter Osnos, *Founder*